Ultrasonic Welding
of Metal Sheets

Ultrasonic Welding of Metal Sheets

Authored by

Susanta Kumar Sahoo,
Mantra Prasad Satpathy

CRC Press
Taylor & Francis Group
Boca Raton London New York

CRC Press is an imprint of the
Taylor & Francis Group, an **informa** business

First edition published 2021
by CRC Press
6000 Broken Sound Parkway NW, Suite 300, Boca Raton, FL 33487-2742
and by CRC Press
2 Park Square, Milton Park, Abingdon, Oxon, OX14 4RN

Library of Congress Cataloging-in-Publication Data

Names: Sahoo, Susanta Kumar, author. | Satpathy, Mantra Prasad, author.
Title: Ultrasonic welding of metal sheets / Susanta Kumar Sahoo, Mantra Prasad Satpathy.
Description: First edition. | Boca Raton : CRC Press, 2021. | Includes bibliographical references and index.
Identifiers: LCCN 2020028009 (print) | LCCN 2020028010 (ebook) | ISBN 9780367265779 (hardback) | ISBN 9780429294051 (ebook)
Subjects: LCSH: Ultrasonic welding. | Sheet-metal work.
Classification: LCC TS228.92 S24 2021 (print) | LCC TS228.92 (ebook) | DDC 671.5/29--dc23
LC record available at https://lccn.loc.gov/2020028009
LC ebook record available at https://lccn.loc.gov/2020028010

Typeset in Times
by KnowledgeWorks Global Ltd.

Contents

Preface

In recent years, a severe energy crisis has become an alarming issue because of the fast draining of nonrenewable energy resources and the unavailability of renewable energy resources to meet human needs. In addition, growing environmental issues such as climate change, toxic CO_2 emissions, and unstable fuel prices have compelled the automotive and aerospace sectors to adopt various strategies for coping with increasing energy consumption and stringent legislative emission regulation. Thus, at this point, any step that helps save energy can be deemed worthwhile in several regards. Out of many approaches, the lightweighting of structures is regarded as one of the most effective ways to achieve the target energy use reductions. Likewise, the consensus on battery-powered electric vehicles is that this technology will become an integral part of society by 2050. Hence, there is an urgent need for an efficient and reliable manufacturing technology which can meet the challenges related to massive weight savings and joining of lithium-ion battery packs.

Ultrasonic spot welding (USW) is a potential means of joining thin, multiple, soft, and conductive metals such as aluminum, copper, brass, magnesium, nickel, steel, and titanium. USW uses high-power ultrasonic vibration to provide relative motion between metal sheets; in response to this vibration, friction and plastic deformation between the intimate metal surfaces causes the temperature to rise. Due to the solid-state nature of this process, the temperature generated is below the melting point of the work materials, so the diffusion of atoms that occurs across the weld interface results in a continuous layer of microweld. Moreover, properties such as low energy consumption, shorter welding time, no use of filler metal, higher efficiency, no shrinkage, and no distortion make USW a desirable emerging technique for joining dissimilar materials. USW could be an alternate method to substitute for conventional fusion welding techniques, as it can reduce the formation of intermetallic compounds (IMCs) with no liquid phase formation. However, the success of USW relies upon various factors such as the selection of work materials and their stacking configuration, weld pressure, weld time/energy, vibration amplitude, and the design of tools and anvil fixtures. The physics of the joining mechanism—from the fundamental concepts, various types of bond failures, vibro-thermographic analyses of tools, the effect of process parameters on the fracture pattern, wake features, flow behavior at plastic deformation zone, and excessive thinning of work material due to acoustic softening—are not well understood. Although experiment-based research is inevitable, there is a trend to supplement such research with computational techniques based on sound theoretical principles.

This book is aimed at presenting the various aspects of the joining of dissimilar metal sheets, ranging from process fundamentals to thorough investigation of metallurgical characteristics that have recently been receiving a lot of attention. The first chapter explains fundamental concepts about the necessity of dissimilar joining in various engineering applications and gives a brief description of the ultrasonic spot welding system. A few illustrations of successful implementation of USW in a variety of manufacturing sectors are highlighted. Chapter 2 gives a detailed explanation of USW, including its complex bonding mechanism and the effects of materials and geometries on weld strength. For better comprehension of the process and to address the critical issues faced in USW, a number of state-of-the-art applications are noted throughout this chapter. Scanning electron microscopy (SEM) with energy dispersive spectroscopy (EDS) also imparts knowledge regarding the evolution of IMCs during the USW process. The design of tools and fixtures plays a vital role in the efficient transmission of ultrasonic energy to the welding zone for producing a satisfactory weld. In this context, Chapter 3 is devoted to providing information regarding many design criteria, such as amplitude uniformity at weld tips, stress distribution, and amplification factor; fatigue and vibro-thermographic analysis of these components are also described. Chapter 4 describes the process of choosing the work material dimensions, and explains the effects of controllable parameters such as frequency of vibration, weld time/energy, weld pressure, and vibration amplitude on performance measures. Chapter 5 attempts to explain the extensive welding experiments that have been done for different dissimilar material combinations. The classifications of distinctive characteristics of joint strength with weld quality highlight the clear pre-eminence of the USW process over other solid-state joining techniques. The numerical simulations have been broadly acknowledged to enhance the accurate prediction of interface temperature and to reveal the significant reduction of the static stress of the material under the influence of ultrasonic energy. Thermo-mechanical modeling is detailed in Chapter 6. Finally, Chapter 7 provides information about several potential research scopes in the USW field.

The authors have strived to provide reliable information on certain aspects of USW of dissimilar metals with various features. The fundamentals and methods in this book can not only enhance the confidence of those attempting to weld dissimilar metal sheets, but may also help solve some of the relevant issues faced by the microelectronics, aviation, and automotive industries. We hope that this book becomes a primary resource for researchers and engineers dealing with the numerous stumbling blocks in dissimilar joining.

S. K. Sahoo
M. P. Satpathy

About the Authors

Dr. Susanta Kumar Sahoo has been a Professor in the Department of Mechanical Engineering at National Institute of Technology, Rourkela, India, since 2008. He was awarded a PhD by Utkal University, Orissa, India, in 2000. He carried out his postdoctoral research in Loughborough University, UK, in 2004. He has been recognized as a leader in the field of solid-state welding for lightweight materials for about the past 10 years. He has 27 years of teaching experience and has published numerous technical papers in national and international journals as well as international conference proceedings. He has guided 15 PhD and 29 MTech. theses and contributed chapters for eight books. Awards conferred upon him include the Young Scientist Award by Odisha Science Academy, the Best Teacher award by the Indian Society of Technical Education, and numerous awards by IE (India) for his valuable contributions to technical education and research. His areas of interest are advanced manufacturing processes, welding technology, metal-forming processes, and advanced machining processes.

Dr. Mantra Prasad Satpathy is currently serving as an Assistant Professor in the School of Mechanical Engineering, KIIT Deemed to be University, Bhubaneswar. He has specialized in the field of materials joining and has four years of teaching and research experience. He completed his PhD on ultrasonic spot welding of dissimilar metals in 2017 at the National Institute of Technology Rourkela, India. He has guided one PG project, and one PhD candidate is working under his supervision in the ultrasonic spot welding field. Dr. Satpathy has received the Sankarsan Jena Memorial Award and the Ganesh Mishra Memorial Award from IEI, India, for his outstanding contributions in the joining field. He has published 21 technical papers in international journals, as well as 23 contributions to international conference proceedings. His areas of interest include simulation and modeling of advanced manufacturing processes, welding technology, metal-forming processes, and advanced machining processes.

1 Introduction

Materials and manufacturing processes have been in use since a long time ago. Over the years, there has been a continuous change in human's need for and choice of materials for various production activities. These activities are sometimes referred to as "ages" when the materials predominantly used had a significant effect on human civilization, such as Stone Age, Copper Age, Iron Age, and (currently) Silicon Age. The contribution of materials in a given society has led to advancement in many sectors, ranging from housing, clothing, transportation, agriculture, medicine, and communication to security and space. The challenges and requirements of the current world are constantly stimulating the invention and advancement of new materials to respond to the complexities of problems. On the one hand, the applications of these new materials let engineers and designers design a myriad of multifaceted structures, but on the other hand, use of these materials also poses a set of unique challenges in terms of integrating a number of components into a single product. These challenges include varying physical, chemical, and mechanical properties, different melting points of materials, and the degree of thinness of sheet materials. For example, mechanical fastening methods, such as self-piercing rivets (SPRs), are usually used to join dissimilar materials in the automotive industry. However, some properties of mechanical fasteners, such as high consumable cost, inconsistent nature, and additional weight, have caused attention to shift toward a robust, fast, and trustworthy welding method. For instance, in the spot welding sector, almost 5,000 weld spots are required to build one car. Numerous welding technologies are present in the current market, but the joining of dissimilar materials is still a challenge.

1.1 FUNDAMENTALS OF WELDING

Most manufacturing processes include some series of operations to produce components having different physical, mechanical, chemical, and dimensional properties. Selection of a proper manufacturing process involves factors such as complexity of the product, production rate, and the related economics. Basically, there are four key manufacturing processes (illustrated in Figure 1.1). Joining is one of the ways to assemble components to attain the desired or preferred shape of the end usable product. *Welding* is the permanent joining process in which metals and nonmetals are joined together, with or without the application of pressure and with or without the addition of filler material [1]. In conventional welding processes, a filler material may be required to assist coalescence. The product created of parts that are joined by welding is called a *weldment*. Welding is commonly applied to join metal parts, but can also be used to join plastics. In contrast, *mechanical joining* is a temporary joining process that uses fasteners such as nuts, bolts, screws, rivets,

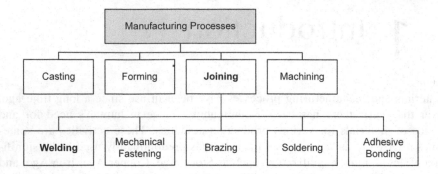

FIGURE 1.1 Basic classification of joining techniques among various manufacturing processes.

and pins. In both brazing and soldering processes, the materials join with each other while a molten filler metal fills the narrow gap between them through the capillary action effect. If the melting point of the filler metal is above 450°C, the process is known as *brazing*. However, if the melting point of the filler metal is below 450°C, the process is termed *soldering*. Likewise, adhesive bonding is a technique of joining two materials using adhesives.

The various advantages of welding techniques include faster integration time, joints that are stronger than the base metal, economy, the light weight of parts, no restrictions as to the site of manufacture, and high joint efficiency. There is a broad spectrum of welding techniques available in the market. However, the selection of a particular welding process depends on the dimension of the components to be welded, the type of material used, the availability of filler materials, the number of units to be produced, economy, accuracy, and reliability. Over a long period of time, it has been shown that certain specific welding techniques are delivering effective outcomes to various particular sectors: automotive sectors use resistance welding, the shipbuilding and heavy engineering industries utilize submerged arc welding, thermite welding is applied to join railway tracks, and aviation industries and nuclear reactors employ tungsten inert gas welding.

1.2 FUSION WELDING TECHNIQUES AND ISSUES

Fusion welding techniques involve the melting of mating surfaces of the base material as well as the filler materials to form weld beads. Thus, heat is an essential component in this process, and it can be supplied from outside in forms of electric arc, gas, resistance heat, and laser beam [2]. Although there are certain advantages of these processes, some factors really make it a challenging task, especially in the case of joining of dissimilar metals. These issues include the generation of residual stress, distortion, and a heat-affected zone (HAZ). Moreover, the mechanical characteristics of the base metal are also severely affected by this strong heating. During the resistance spot welding (RSW) process, some amount of contact pressure (along with the electric resistance) is required to produce coalescence at the faying surface. RSW is widely used in the automobile, heavy engineering, and electronics sectors due to its fast, automated, and reliable nature. However, this technique encounters many hurdles during the welding of materials of high electrical and thermal conductivities with dissimilar properties [3]. Moreover, the process requires high energy (i.e., 50–100 kilojoules [kJ] per single weld spot) [4]. This process is more prone to form intermetallic compounds during the welding process; create a large HAZ; and suffer distortion, cracks, residual stress, and short electrode life. Figure 1.2 portrays a typical interface failure in RSW.

Laser welding is a type of noncontact technique that joins metal sheets by melting them down in a few seconds. Despite its ability to join multiple layers of sheets by the laser beam, this welding process also has certain drawbacks, such as poor metallurgical affinity, porosity, large intermetallic compound formation, high initial set-up cost, and sensitivity to cracks [6]. The various defects that may be produced during laser welding are depicted in Figure 1.3.

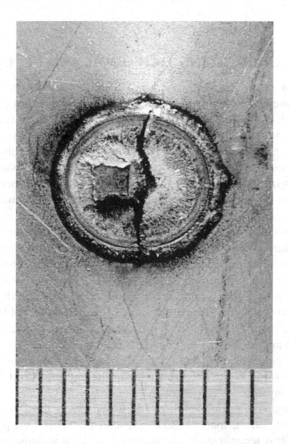

FIGURE 1.2 Interfacial weld defect showing a prominent crack from the RSW process [5].

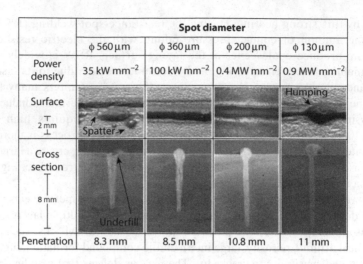

	Spot diameter			
	φ 560 μm	φ 360 μm	φ 200 μm	φ 130 μm
Power density	35 kW mm^{-2}	100 kW mm^{-2}	0.4 MW mm^{-2}	0.9 MW mm^{-2}
Surface ⊤ 2 mm ⊥	Spatter			Humping
Cross section ⊤ 8 mm ⊥	Underfill			
Penetration	8.3 mm	8.5 mm	10.8 mm	11 mm

FIGURE 1.3   Surface features and defects at cross-sections of weld beads at various laser spot diameters [7].

1.3   SOLID-STATE WELDING TECHNIQUES AND ISSUES

Solid-state welding is a kind of process in which the faying surfaces of the base material get joined without melting. Pressure must be applied to get a good joint by forming a thick metallic bond between the workpieces. It is worth noting that the temperature involved in all solid-state welding techniques is below the melting point of the base materials. This technique utilizes the combination of heat and pressure for a particular time period to overcome hurdles such as porosity, cracks, and slag inclusions faced by the conventional welding process. Solid-state welding includes the pressure, forge, friction, explosive, diffusion, and ultrasonic welding processes. The basic principle of these processes is the atomic diffusion between the two materials, which will occur when the interatomic distance is below 10 Å. Thus, these types of welding methods require clean and perfectly smooth surfaces. The major hindrances of these techniques are the formation of an oxide layer and the presence of oils and other contaminants. These factors make the solid-state welding process a little more complicated than other welding methods. However, the distinct advantages of solid-state welding techniques include no requirement for filler material; lower value of residual stress, distortion, and HAZ; and no change in the mechanical properties of the base material. However, the expensive equipment, long specimen preparation time, complicated joint design, and nonconductive properties of the material limit the use of this technique in many cases.

Although friction stir spot welding (FSSW) is widely used in the automobile manufacturing sector, the longer weld time and exit hole after the process are limiting factors of this technique [8, 9]. Another solid-state welding process that has not garnered much attention commercially is the ultrasonic spot welding (USW)

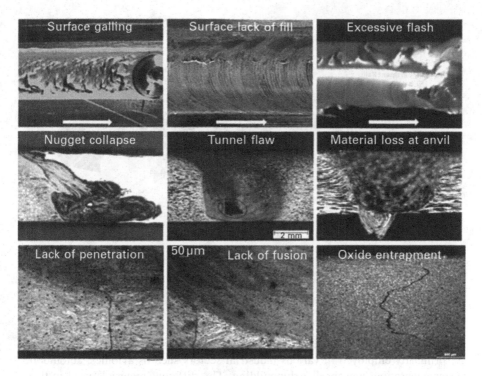

FIGURE 1.4 Various types of flaws in the FSSW process [12].

technique. It is more effective than FSSW, as the energy is generated at the weld interface in this process, rather than only at the upper surface of the specimen as in FSSW [10]. Furthermore, during the production of 1,000 weld joints, the USW process expends a weld energy of 0.3 kWh, whereas RSW and FSSW utilize 20 kWh and 2 kWh, respectively [11]. USW typically requires less than 0.5 seconds weld cycle time and produces higher-quality and stronger joints in comparison with FSSW technique on the basis of the same weld nugget area. The various flaws in FSSW are illustrated in Figure 1.4.

1.4 NECESSITY OF DISSIMILAR JOINTS IN ENGINEERING STRUCTURES

Dissimilar material joining concerns arise because of material combinations that are usually challenging to join due to substantial dissimilarities in physical, thermal, and mechanical properties (ductility, fatigue strength, elastic modulus, etc.), as well as chemical compositions. Additionally, issues related to the galvanic corrosion and different thermal expansion of dissimilar metals must be carefully handled. The structures created through this joining process are referred as *multi-material hybrid structures*, and they are optimally designed to serve a specific purpose [13]. Meanwhile, the demand for these types of lightweight, high-performance-with-great-functionality

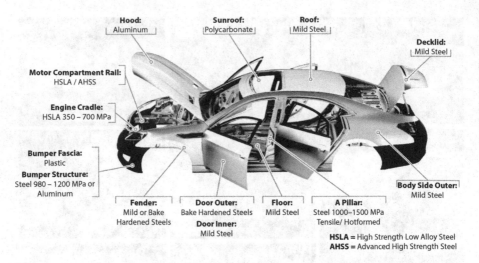

FIGURE 1.5 Type of material used in 14 major automotive components of the body structure [14].

structures is increasing day by day. These types of multi-material hybrid structures are primarily employed in various industrial sectors such as avionics, automobile, power generation, and marine shipbuilding. The different materials used in these structures cannot be selected at random; rather, a systematic approach is required to ensure that the selected materials are compatible with each other. Furthermore, a new manufacturing system may be required to retain the weight-to-strength ratio, production cost, and environmental and recyclability features of these structures. Figure 1.5 displays the various materials needed in the 14 major components for production of an automobile.

The modern automobile sector has adopted different materials for several components, such as high-strength low-alloy (HSLA) steel for body and chassis structures, aluminum for bumper and hood, mild steel for roof and deck, and composite sheets for sunroof and panels. The European superlight car projects [15] utilized these types of multi-material structures as illustrated in Figure 1.6.

Usually, the aeronautics sector exploits aluminum and steel materials to manufacture various components. The preference for these materials is due to the facts that steel is a cheap, easy-to-work, and robust metal, and aluminum is lightweight and resistant to both corrosion and rust. Moreover, this combination improves the outcome in many ways. Figure 1.7 shows a typical aircraft component fabricated by dissimilar welding of aluminum and steel.

Of course, weight is a crucial factor in the aviation industries. Engineers have always put effort into enhancing the weight-to-strength ratio while minimizing the operating costs of their structures. Thus, the multi-material design concept plays a significant role in this context, and it also simplifies the aircraft maintenance service. Lightweight metals, such as aluminum and magnesium, and composite materials are the economically viable alternatives that can last for

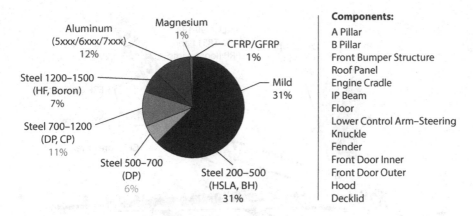

FIGURE 1.6 Various materials used in major automotive structural components [14].

a long time due to their substantial mechanical strength and improved physical properties in various harsh environments. The multiple materials utilized in crucial structural components of an Airbus A380 are illustrated in Figure 1.8. Likewise, the Boeing 787 Dreamliner also employs 20% aluminum alloys, 15% titanium alloys, 10% steel sheets, and 5% other metals in the manufacturing of the airframe structure [17].

The battery electronics industries use dissimilar-metal welding procedures to join the tabs with the electrode leads. However, it is not easy to combine these multiple highly conductive and thin layers with exceptional reliability. Furthermore, the performance of a battery is significantly affected as it is exposed to the critical driving

FIGURE 1.7 A welded structure composed of aluminum and steel, used in the aircraft industry [16].

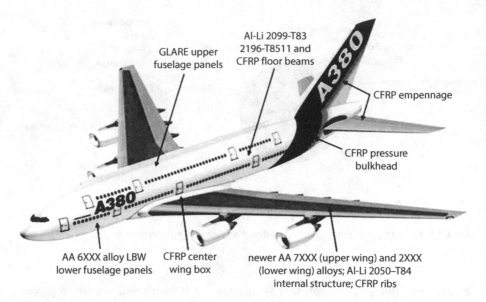

FIGURE 1.8   Different materials distributions for the Airbus A380 [18].

environment, elevated temperature, and crash. Figure 1.9 displays dissimilar joining in the battery manufacturing process.

The propulsion system in an aircraft requires advanced materials, design, and fabrication techniques. These high-performance materials include titanium and nickel-based superalloys, shape-memory alloys, carbon-fiber composites, and ceramics. These materials can be integrated in the aircraft engine, including metal-to-metal and metal-to-ceramic materials (Figure 1.10).

Each welding technology has its own strengths and limitations that affect the joining of dissimilar materials. However, to produce a satisfactory weld, the product

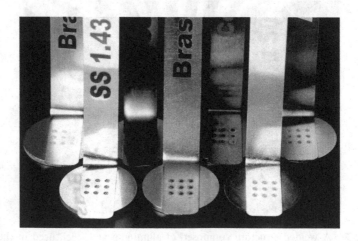

FIGURE 1.9   Dissimilar metals welded to form batteries [19].

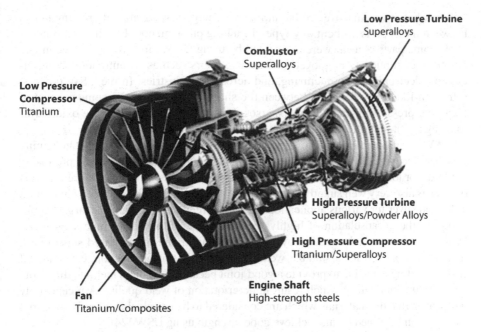

FIGURE 1.10 Various metals utilized in the jet-propulsion engine of an aircraft [20].

design and the design of the joining process are the crucial factors. According to Messler [21], a good weld can be achieved through minimization of the atomic distance between the two dissimilar metals. However, it is challenging to control this distance during the welding process.

1.5 ULTRASONIC SPOT WELDING

Sound waves consist of a sequence of compression and rarefaction through which acoustic energy propagates. The human ear can typically detect sounds in the frequency range from 20 to 20 kHz, which is called the *audible range*. The upper and lower limits of this range are called *ultrasound* and *infrasound range*. Nowadays, the applications of ultrasonics are receiving lots of attention in the manufacturing industries as well as in medical sciences. According to the rate of energy transfer, ultrasonics is divided into two categories: low-power and high-power ultrasonics. Generally, in nondestructive testing and medical diagnostic applications, low-power ultrasonics are used, with frequencies of more than 100 kHz and power intensity ranging from 0.1 to 0.5 W/cm². Similarly, high-power ultrasonics are applied in welding, cutting, and drilling operations, with equipment whose frequency range varies from 20 to 100 kHz with power intensity greater than 100 W/cm² [22].

In the 1930s, ultrasonic energy was applied to refine the grain size of the molten metals during the soldering process. Ultrasonic spot welding was first demonstrated in the early 1950s when Aeroprojects® (currently known as Sonobond® Corporation) [23] applied low-frequency vibration to obtain a cold weld at the resistance spot weld region. However, the real and full range of applications of USW started in the late

1960s, when it was discovered that ultrasonic energy was capable of creating bonds between metal parts without any type of melting phenomenon. For the first time, in 1969, some parts of a car were assembled by using USW. Since then, this technique has been continuously employed in various sectors such as an automobile, electrical and electronics, manufacturing, and aerospace industries. In the USW process, a friction-like relative motion between the sheets will occur under some amount of clamping pressure. This motion disperses the surface asperities and the oxide layer to bring about metal-to-metal contact.

USW is considered to be a novel and innovative technique for the solid-state joining process. Recently, USW has drawn significant attention as one of the most promising methods for joining dissimilar metals such as plates, sheets, wires, etc. Moreover, this process is also very well suited for the joining of soft metals such as copper, aluminum, gold, and silver for various applications. USW currently finds one of the largest application in the encapsulation of highly sensitive chemical or pyrotechnic substances. With the advent of advanced alloys such as titanium and nickel-based superalloys, there is also a need for a suitable welding process. In this application, conventional fusion welding is unable to provide a good joint because it requires melting, diffusion, and solidification, all of which result in deterioration of weld quality. A recent study has proven that the materials which are considered to be unweldable by resistance spot welding can be joined by and achieve good strength using USW [24].

Typically, ultrasonic welding is divided into two categories according to its applications: (1) ultrasonic metal welding and (2) ultrasonic plastic welding. The significant difference between these two systems is that in the case of ultrasonic metal welding, the vibration is parallel to the surface of the weld materials, whereas in ultrasonic plastic welding the vibration is perpendicular to the surface. Both types of systems consist of five standard subsystems: (1) ultrasonic frequency generator, (2) piezoelectric transducer, (3) booster, (4) horn/sonotrode, and (5) anvil/fixture. The various components of a typical ultrasonic metal welding system are presented in Figure 1.11.

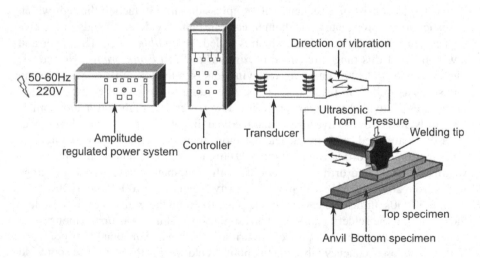

FIGURE 1.11   Schematic diagram of a USW system showing different parts.

1.6 BRIEF DESCRIPTION OF ULTRASONIC SPOT WELDING EQUIPMENT

In the automotive and microelectronic industries, several types of USW machines are utilized depending on the kind of weld needed, the geometry and varieties of metals, and so on. Typically, all devices employ a transverse mode of vibration to weld metal sheets. Despite the availability of several ultrasonic welding methods (such as seam, ring, line, and torsion), spot welders are widely employed because of their simplicity of operation, superb weld quality, low capital cost, and robust nature. Ultrasonic spot welders come in two configurations: the lateral drive system and the wedge-reed system. These two types of systems have different shapes and applications, but the principle of the deformation in the weld area by these vibrational mechanisms is the same.

1.6.1 LATERAL DRIVE SPOT WELDING SYSTEM

A lateral drive spot welding system is comprised of an ultrasonic frequency generator with a power amplifier, transducer, booster, sonotrode with weldable tips, and anvil, as shown in Figure 1.12. Initially, the 50 Hz frequency of a standard

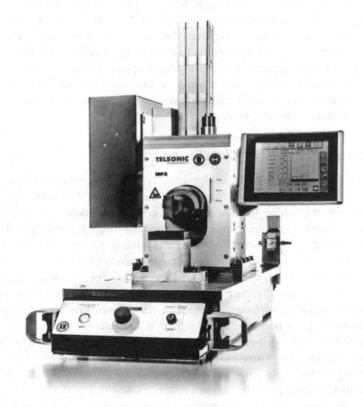

FIGURE 1.12 Lateral drive ultrasonic spot welder [25].

alternating current (AC) supply is converted to high-frequency power (15–70 kHz) with the help of the ultrasonic frequency generator, and its strength is increased by the power amplifier. The piezoelectric transducer/converter is made up of lead zirconate titanate (PbZnTi) crystals/ceramic rings; when a high-frequency voltage is applied to these rings, a large mechanical displacement is observed. The nature of this displacement is longitudinal, and it resonates at a frequency governed by the piezoelectric ceramics (front and rear drivers). At the nodal position, the minimum vibration amplitude can be experienced, so it is designed to be positioned at the flange. Meanwhile, a higher vibrational amplitude can be attained at the front driver end. However, this amplitude is quite a bit less (10–30 μm) than is needed to perform the welding operation. The booster subsequently amplifies the transducer's vibration amplitude depending upon its geometries, and it also serves as mounting for the entire welding stack. The horn or mechanical amplifier again increases these values up to a level that will be sufficient for welding. Another role of the horn is to act as a tool during the welding process. As there is direct contact between the tool and the workpiece during welding, wear and tear will happen to the horn. Thus, some sonotrodes are designed with replaceable tips. Nevertheless, most sonotrodes are designed to act as a single unit in order to reduce energy loss. Moreover, the materials for the horn and booster should be chosen so that these parts can withstand a high wear rate, corrosion, and fatigue loads. Thus, titanium and tool steels are commonly used as booster and horn materials. In this type of welding system, the horn is placed parallel to the workpiece. An anvil is used to support the fixture and to hold the bottom specimen firmly. Consequently, a sufficient relative motion between the sheets is observed without any slippage. Therefore, the ultrasonic energy is transferred to the weld coupon in a transverse manner. The specimens, usually thin metal sheets, are overlapped on each other with the bottom specimen firmly attached to the anvil surface.

1.6.2   WEDGE-REED SPOT WELDING SYSTEM

Like the lateral drive system, the wedge-reed system also consists of five components: a generator, transducer, wedge, reed, and anvil, as shown in Figure 1.13. In this system, the wedge has the same function as a booster. The ultrasonic energy is transferred from the transducer to the reed through the wedge. The reed vibrates in a bending mode and produces transverse vibrations at the welding tip. Generally, the wedge is attached to the reed by means of welding or brazing to avoid any type of loss during the transfer of vibration energy. However, in some situations, the anvil also vibrates and resonates out of phase to increase the relative motion between the surfaces. Meanwhile, the transducer does not receive any direct resistance from the weld spot; rather, it only drives the reed. Thus, the parameters at the weld spot cannot be controlled accurately. This sort of wedge-reed system works at low vibration amplitude with a high clamping force, capable of producing high-quality joints between high-strength alloy sheets. Although the vibrational amplitude in this system is about one-third of the clamping force, it is three times higher than the amplitude produced by the lateral drive system

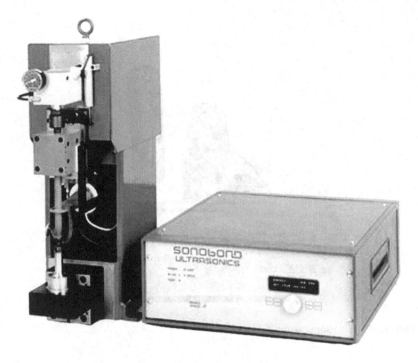

FIGURE 1.13 Wedge-reed ultrasonic metal spot welder [26].

running at a similar power level. Likewise, the impedance of the wedge-reed system is about nine times higher than that of the lateral drive system. Overall, the wedge-reed system outperforms the lateral drive system at similar or higher power levels due to its superior power at the weld zone. Furthermore, in tin-coated wires it also provides weld joints of high quality and durability due to its shear mode vibration.

1.6.3 OTHER ULTRASONIC VIBRATION-ASSISTED WELDING PROCESSES

Aside from the standard spot welding equipment, ultrasonics can be applied for wire bonding, seam welding, torsion welding, and ultrasonic additive manufacturing. Wire bonding, specifically a small-scale ultrasonic welding technique, is used to connect microelectronic parts. The second type of weld achieved through ultrasonics is the seam weld, shown in Figure 1.14. It has a rotating lateral drive transducer, which is attached to a circular disk-type sonotrode for producing the continuous weld. The bearings are attached in this system on both the sides to provide a turning moment to the sonotrode. These types of sonotrodes are made up of hardened steel, which is extremely resistant to wear during cut-and-seal operations. In this kind of system, the anvil may move according to the turning of the sonotrode. These types of welding machines are extensively used in joining aluminum with copper. Ultrasonic seam welders are also highly capable of producing effective circumferential welds

FIGURE 1.14 Ultrasonic seam welder [27].

of round specimens with any diameter. Currently, the textile industries are replacing sewing machines with this type of ultrasonic seam welding machines for fabrics made of thermoplastic fibers.

Another means of achieving a good ultrasonic weld is the use of an ultrasonic torsional welder (UTW), shown in Figure 1.15. This system was first developed in 2002, and at that time it was employed to seal nonferrous metals such as cylindrical lead-acid batteries, automotive seat belt sensors, and air bags. Typically, the automotive industry utilized this seal welding technique to insert 50,000 sensors in seat belts and airbags daily. This practice also finds a suitable place in plastic welding applications such as camshaft and sunlight sensors, medical instruments, and foil valves. However, from 2009 onward, this technique has been applied to weld metal parts such as battery tab connections, capacitor connectors, spray valves, and wire harnesses. The ultrasonic generator used in this UTW procedure supplies 20 kHz of sinusoidal frequency to one or more converters. The piezoelectric discs present in these converters contract and expand under these positive and negative voltages oscillations. The ultrasonic transducers connected to the booster and sonotrode produce push-pull or torsional vibration. This kind of welding system can provide a circular-pattern weld spot on the workpiece. The torsional oscillations reach a maximum at the tip of the sonotrode.

These multiple converters accelerate the joining process; thus, it is faster than the conventional ultrasonic spot welding process. In the UTW process, the sonotrode need not be located at the center of the weld spot. However, all the components of this system should resonate at the same frequency to deliver maximum energy to the weld zone. The basic difference between a traditional ultrasonic spot welder and

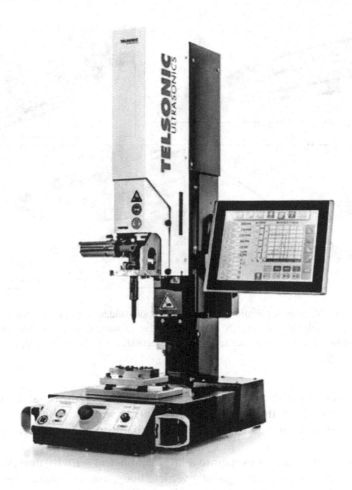

FIGURE 1.15   Ultrasonic torsional welder [28].

UTW is the mode of transmission of ultrasonic vibration to the weld zone. Usually, the ultrasonic spot welder applies the vibration at the weld interface in a transverse manner, whereas UTW applies the transverse vibration with a twisting moment. Thus, the UTW produces stronger welds at the molecular level. It can be used to weld large parts and also produces narrower welds. Thus, hybrid electric vehicle makers have used this technique to join huge amounts of wire terminals.

1.7   SELECTED ILLUSTRATIVE ULTRASONIC SPOT WELDING APPLICATIONS

The superior performance of ultrasonically welded joints, such as high mechanical strength, high corrosion resistance, better electrical conductivity, and better fatigue properties, enables ultrasonic welding to be used in many sectors,

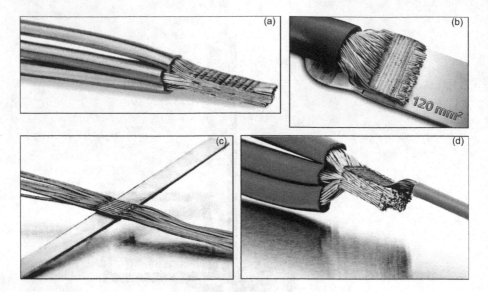

FIGURE 1.16 Various application of ultrasonic spot welded wire connections: (a) multi-stranded copper wire connection, (b) aluminum cable to aluminum conductor rail connection, (c) enameled wires to copper terminal connection, (d) heavy-duty copper-aluminum connection [29].

including aerospace, automotive, microelectronics, medical, and consumer goods. Illustrations of the modern-day parts made with ultrasonic spot welding appear throughout this section to show the reliability and productivity of this process.

The automotive industries require a large number of wire harnesses to transmit the electrical signals throughout a vehicle body, and such harnesses are created by the consolidation of multiple solid wires of different thicknesses. Figure 1.16(a) shows the compaction of multiple strands of copper wires of various sections for an electrical wire harness used in an automobile. Due to the comparatively high price of copper, aluminum wires are increasingly being used as transmitters of high voltage. Figure 1.16(b) shows the joining of an aluminum cable to the aluminum conductor rails that are typically used to conduct power from the battery pack to the engine compartment. This connection quality is so high that it offers low electrical contact resistance and minimizes conduction loss. USW has been successfully applied to weld enameled wires to a copper terminal without first removing the enamel coating (Figure 1.16(c)). This connection shows excellent strength, and electrical conductivity is not affected. Figure 1.16(d) illustrates heavy-duty multi-stranded aluminum and copper connected to each other in a wire harness.

MAK terminals (Figure 1.17(a)) are employed in vehicle construction for high-current applications. This terminal is made up of copper alloy, and it is covered by a stainless steel casing. This type of terminal is extremely robust, highly electrically

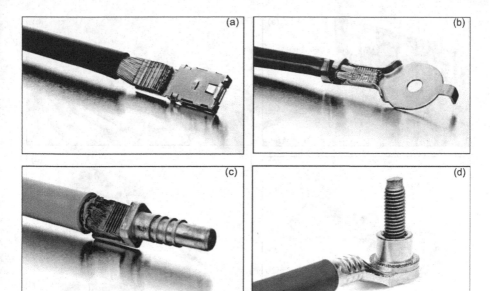

FIGURE 1.17 Applications of ultrasonic spot welded wires to various terminals: (a) copper cables to MAK terminal, (b) multiple aluminum wires to copper eyelet terminal, (c) high-flex copper contact, (d) torsional welded bolt on conductor rail [29].

conductive, and very reliable. The grounding connections of vehicles are attached to certain points of the gear by using an eyelet terminal (Figure 1.17(b)). Figure 1.17(c) shows a high-voltage-carrying copper cable. In vehicles, aluminum conductor rails have gradually come to be used due to their lower cost, lower weight, and good electrical conductivity (Figure 1.17(d)). This type of insulated aluminum conductor rail, with a specific type of terminal, must be equipped with screws for useful contact purposes. These screws are joined to the conductor rails by ultrasonic torsional welding.

Likewise, bus bars, terminators, contact assemblies, and sensor terminals are all examples of connections produced by ultrasonic welding, and these are now successfully used in the electronics and automotive industries (Figure 1.18). This welding technique offers low electric contact resistance and avoids any types of cracks on the ceramic board. In fact, ultrasonic micro-bonding is also widely used for microelectronics interconnections.

Battery and fuel cell manufacturers also use ultrasonics to make joints between foils to thin copper, nickel, or aluminum tabs. Therefore, the USW technique is extensively employed in the construction of electric and hybrid vehicles, because it is capable of joining multiple layers of foils with the tabs located inside of lithium-ion batteries. Moreover, the ultrasonic weld has been extensively used in electrical industries for joining electrical motor coils, field coils, capacitors, and transformers

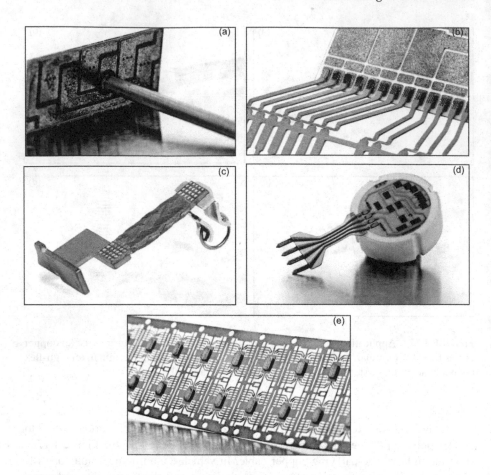

FIGURE 1.18 Various applications of ultrasonic spot welding in microelectronics: (a) ceramic printed circuit board with copper conductor, (b) electrical connections on insulated gate bipolar transistor (IGBT) modules, (c) switch assembly with braided copper wire, (d) flat cable connector with printed circuit board (PCB) on ceramic substrate, (e) perforated copper contact stud on sensitive copper strip [30].

(Figure 1.19). USW is a versatile technique which can be applied in any condition—including in water or in a vacuum—to produce a satisfactory weld [31]. Packaging industries are also using ultrasonic seam, torsion, or spot welding systems for sealing purposes. Seam welding is used to seal to cooking pouches and food foils. Similarly, the torsion welding system is used to seal the cylindrical units that contain reactive and hazardous materials.

Another advantage of USW is its energy use. It has been shown that the conventional welding process (such as resistance welding) consumes 10% more energy than does USW [33]. Recently, USMW has been tried for joining thin sheets of aluminum with other lightweight materials in structural automotive and aerospace

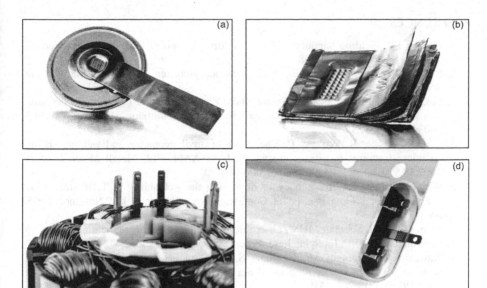

FIGURE 1.19 Applications of ultrasonic spot welding technique in battery and electrical sectors: (a) copper jumper attached to aluminum battery cap, (b) copper foils welded to nickel-coated copper bus bar in lithium-ion battery, (c) enameled copper winding wires connected to stator in electric motor, (d) nickel-plated copper strip attached to aluminum housing in capacitor [32].

applications [34]. This process is being used to assemble secondary aircraft structures such as helicopter access doors., Doors made with this process were able to sustain loads five to ten times the design load without weld failure, as shown in airload tests described by the American welding society (Figure 1.20) [35].

FIGURE 1.20 Ultrasonic spot welded helicopter access door [23].

REFERENCES

1. Connor LP. Welding handbook. Vol. I: Welding technology. Miami, FL: American Welding Society; 1987, p. 638.
2. Phillips DH. Welding engineering: An introduction. Hoboken, New Jersey: John Wiley & Sons; 2016.
3. Fukumoto S, Lum I, Biro E, Boomer DR, Zhou Y. Effects of electrode degradation on electrode life in resistance spot welding of aluminum alloy 5182. Weld J 2003;82:307S–12S.
4. Jahn R, Cooper R, Wilkosz D. The effect of anvil geometry and welding energy on microstructures in ultrasonic spot welds of AA6111-T4. Metall Mater Trans A 2007;38:570–83. doi: 10.1007/s11661-006-9087-0.
5. Shi G, Westgate SA. Techniques for improving the weldability of TRIP steel using resistance spot welding. Pap. 1st Int. Conf. High Strength Steel. UK; 2 November 2005, pp. 1–13.
6. Esser G, Mys I, Schmidt MHM. Laser micro welding of copper and aluminium using filler materials. Fifth Int Symp Laser Precis Microfabr 2004;5662:337–43.
7. Kawahito Y, Mizutani M, Katayama S. High quality welding of stainless steel with 10 kW high power fibre laser. Sci Technol Weld Join 2009;14:288–94. doi: 10.1179/136217108X372531.
8. Chowdhury SH, Chen DL, Bhole SD, Cao X, Wanjara P. Lap shear strength and fatigue life of friction stir spot welded AZ31 magnesium and 5754 aluminum alloys. Mater Sci Eng A 2012;556:500–9.
9. Prangnell P, Haddadi F, Chen YC. Ultrasonic spot welding of aluminium to steel for automotive applications—microstructure and optimisation. Mater Sci Technol 2011;27:617–24.
10. Bakavos D, Prangnell PB. Effect of reduced or zero pin length and anvil insulation on friction stir spot welding thin gauge 6111 automotive sheet. Sci Technol Weld Join 2009;14:443–56.
11. Bakavos D, Prangnell PB. Mechanisms of joint and microstructure formation in high power ultrasonic spot welding 6111 aluminium automotive sheet. Mater Sci Eng A 2010;527:6320–34.
12. Zettler R, Vugrin T, Schmücker M. Effects and defects of friction stir welds. In: Lohwasser D., Chen Z. (eds). Frict. Stir Weld., Elsevier, Cambridge, UK; 2010, pp. 245–76.
13. Sirisalee P, Ashby MF, Parks GT, John Clarkson P. Multi-criteria material selection of monolithic and multi-materials in engineering design. Adv Eng Mater 2006;8:48–56.
14. Center for Automotive Research. Technology roadmaps: intelligent mobility technology; materials and manufacturing processes, and light duty vehicle propulsion. ASHA Leader 2017;22:20–22. doi: 10.1044/leader.ppl.22062017.20.
15. SLC Consortium. Super LIGHT-CAR Introduction 2014. http://www.superlightcar.com/public/index.php (accessed February 1, 2019).
16. SPI Lasers Limited. Dissimilar Metal Welding 2018. https://www.spilasers.com/application-welding/dissimilar-metal-welding/ (accessed February 5, 2019).
17. Wikipedia Contributors. Boeing 787 Dreamliner 2014. https://en.wikipedia.org/w/index.php?title=Boeing_787_Dreamliner&oldid=619630395 (accessed February 11, 2019).
18. Wanhill RJH. Carbon fibre polymer matrix structural composites. In: Eswara Prasad N, Wanhill RJH (eds). Aerosp. Mater. Mater. Technol. Singapore: Springer; 2017, pp. 309–41.
19. SPI Lasers Limited. Dissimilar metal welding in the electronics industry 2018. https://www.spilasers.com/application-welding/dissimilar-metal-welding-electronics-industry/ (accessed February 11, 2019).

20. Hassey P. Mission critical metallics. ATI 2010. https://www.sec.gov/Archives/edgar/data/814250/000095012310110284/l41250exv99w1.htm (accessed February 12, 2019).
21. Messler RW. Joining of materials and structures: From pragmatic process to enabling technology. NY: Butterworth-Heinemann; 2004.
22. McCulloch E. Experimental and finite element modelling of ultrasonic cutting of food. PhD thesis, Glasgow: University of Glasgow; 2008.
23. Graff K. Ultrasonic metal welding. In: Ahmed N, ed. New Dev. Adv. Weld., Cambridge: Woodhead; 2005, pp. 241–69.
24. Graff K. Development of advanced joining processes for AHSS and UHSS. Columbus, OH: Edison Welding Institute (EWI); 2007.
25. Telsonic Ultrasonics. MPX-ultrasonics-welding-system 2019. https://www.telsonic.com/en/products/mpx-ultrasonics-welding-system/ (accessed March 12, 2019).
26. Sonobond Ultrasonics. SonoWeld 1600 ultrasonic metal spot welder 2015. https://www.sonobondultrasonics.com/welders-bonders/metal-welders/sonoweld-1600-spot-welder (accessed March 12, 2019).
27. Sonics. Seam welders 2019. https://www.sonics.com/metal-welding/products/seam-welders/ (accessed March 2, 2019).
28. Telsonic Ultrasonics. Torsional ultrasonic welding: a gentle yet powerful process 2014. https://www.telsonic.com/en/news-ultrasonics/detail/article/torsional-ultrasonic-welding-a-gentle-yet-powerful-process/ (accessed March 2, 2019).
29. Telsonic Ultrasonics. Applications: Automotive 2017. https://www.telsonic.com/en/application-finder/#metal_welding (accessed March 12, 2019).
30. Telsonic Ultrasonics. Applications: Electronics 2017. https://www.telsonic.com/en/application-finder/#metal_welding (accessed March 12, 2019).
31. Matheny M. Ultrasonic metal welding foils to tabs for lithium-ion battery cells. EWI Summ Rep SR1301 2012. https://ewi.org/ultrasonic-metal-welding-for-lithium-ion-battery-cells-2/
32. Telsonic Ultrasonics. Applications: Battery 2017. https://www.telsonic.com/en/application-finder/#metal_welding (accessed March 12, 2019).
33. Bakavos D, Prangnell PB. Mechanisms of joint and microstructure formation in high power ultrasonic spot welding 6111 aluminium automotive sheet. Mater Sci Eng A 2010;527:6320–34. doi: 10.1016/j.msea.2010.06.038.
34. Froes FH. Advanced metals for aerospace and automotive use. Mater Sci Eng A 1994;184:119–33.
35. O'Brien RL. Appendix 1: history of welding and cutting. Jefferson's weld encycl 18th edn. In: O'Brien (ed). Miami, FL: American Welding Society (AWS); 1997.

2 Fundamentals of Ultrasonic Spot Welding

Dissimilar materials can be joined successfully using new and innovative industrial automotive welding techniques that require less energy, reduce cycle time and operating costs, and improve quality, among other advantages. Conventional fusion welding techniques operate above the melting point of the base material. The mechanical properties and microstructural characteristics of the joint are significantly affected by this high heat input and rapid solidification of molten metal. As a result, cracks due to shrinkage, distortion, and residual stress defects occur. These issues can be solved by the introduction of solid-state joining techniques that include the friction phenomenon. One of these solid-state joining techniques, ultrasonic spot welding (USW), is a new and emerging concept used in industry over the past 20 years and serving manufacturing sectors such as aviation, medical, microelectronics, automotive, and much more. USW is a clean and reliable technique in which the welding is done with high energy. No flux or filler metal is needed, tool life is longer, and it takes a very short time (less than one second) to weld materials in a perfectly controllable environment with better efficiency and a higher mass production rate.

2.1 OVERVIEW OF THE ULTRASONIC METAL WELDING PROCESS

According to the definition from the American Welding Society (AWS), *welding* is the process of joining two or more metal parts by localized coalescence across their interface [1]. Thus, to create a weld, coalescence must be achieved by bringing the specimens so close that the distance between the atoms is equal to crystal lattice spacing [2]. Recently, the principle of joint formation in USW has been elaborated in a number of articles. It is commonly understood from these reports that the USW technique involves three different stages.

Stage I: The weld surfaces are brought close by the application of normal force through sonotrode and anvil (due to the clamping pressure) along with shearing force (due to transverse vibration of parts) in a USW process. The surfaces of the metals may have oxide coatings as well as contaminants such as moisture and lubricants, which prevent proper metal-to-metal contact. Furthermore, the surfaces of the specimens may have a lot of peaks and valleys depending upon their roughness. Thus, the initial numbers of contact points depend on the surface roughness and the normal force applied to it.

Stage II: The ultrasonic vibration is started along with normal force, and in most cases the top specimen moves relative to the bottom one. Due to this compressive force and frictional relative motion at the weld interface, the initiation of welding commences with generation of atomic dislocations. Thus, chemical bonds develop between the surfaces of weld metals at a distance of approximately 4 to 5 nm, and the interchange of electrons takes place between the two surfaces. This leads to the plastic deformation and shearing of the interfering asperities from the weld zone, resulting in an atomically clean weld surface.

Stage III: This stage, which is crucial to the production of a strong joint between the surfaces, includes the diffusion of atoms across the bonding line on a microscopic scale. The weld interface in this stage is subjected to severe plastic deformation, and this leads to the splitting of grains and crystals to form smaller subgrains. Thus, small weld zones (called *microwelds*) [3] are evolved. In the course of time, the vibration increases the microweld areas, and finally the weld area is completely filled with weld zones. The area just below the plastically deformed zone is exposed to the elevated ultrasonic energy and thus residual stress is formed in this area. However, this residual stress is eventually relaxed in this zone due to the quick generation of high temperature that causes the atoms to alter their functional positions in the crystal lattice. Figure 2.1 shows a schematic diagram of the various stages of the USW process.

The nature of the bond that is developed between the interfaces in a USW process is a solid-state type. This type of bond is achieved without melting and fusion of the workpieces. Nevertheless, the plastic deformation in the weld zone results in a noticeable rise of temperature in varying welding conditions. It is well known that the yield strengths of most metals are sensitive to the increase of temperature. Hence, temperature rise reduces the yield strength of the material significantly and, in turn, it further enhances the plastic deformation and flow of materials in the weld zone.

In USW, the amount of plastic deformation and yielding of the top sheet depend on the process parameters as well as specimen-related properties. Process parameters

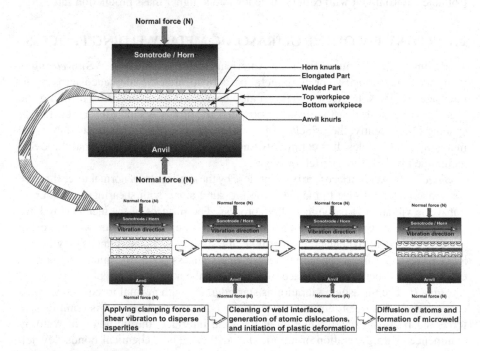

FIGURE 2.1 Schematic diagram of bond formation at weld interface during the USW process.

include vibrational amplitude, weld pressure, and weld time. Similarly, specimen parameters consist of the thermal conductivity, hardness, and roughness of the sheet being welded [4–6]. As USW is a complex process, the control of process parameters is difficult. Likewise, the thickness of the workpiece and the weld tip size also determine the quality of the weld. If the material is thick, then high power is required, along with a large weld tip, to transfer the ultrasonic energy efficiently to the weld zone. However, higher force may hinder the relative motion between the sheets. Thus, before finalizing the parameters, a lot of trials and experiments are required to establish a suitable range.

2.2 WELDABLE MATERIALS AND THEIR GEOMETRIES

Ultrasonic spot welding, in most cases, allows the joining of similar and dissimilar materials of thinner dimensions. This process may also be applied to materials of dissimilar thickness as well as to multiple-layer sheets [7]. In this regard, material properties play a vital role, which require a proper investigation to determine whether they are amenable to USW. The AWS introduced the term *weldability*, which is defined as the ease with which the materials may be joined to meet the service conditions. Over the years, a lot of materials have been considered for their weldability in general and for ultrasonic welding in particular. Their appropriateness depends primarily on the material hardness and crystal structure. Typically, ultrasonic weldability of the material decreases with the increase of hardness. We know that the crystal structure divides metals into three categories: face-centered cubic (FCC), body-centered cubic (BCC), and hexagonal close-packed (HCP). Metals like Al, Cu, Au, Ag, and Pt with FCC structure are the most weldable materials. Metals such as Mg, Ti, and Zn with HCP structure have limited weldability. Similarly, metals with BCC structure, such as Fe, Cr, Mo, and W, have weldability that lies in between those of FCC and HCP materials. A summary of weldability chart, based on data as obtained from experimental observations, is presented in Figure 2.2. The shaded portions are supposed to represent the combinations that can be welded and the blank portions represent the combinations that cannot be welded or have not been attempted yet. Figure 2.2 shows that aluminum is exceptionally weldable and can be joined to almost all metals. Other easily weldable materials are copper alloys and various precious metals. In contrast, iron and refractory materials such as molybdenum and tungsten can be welded only if the materials are thin [8]. Apart from joining of the metals, ultrasonic welding experiments have also been conducted on 1.5-mm-thick SiCp/2009Al-O and SiCp/2009Al-T6 composites to reveal the welding feasibility of these composites [9].

In an industrial welding scenario, there are various factors such as specimen thickness, specimen dimensions, weld zone positioning, vibrational attributes, material properties, and environmental conditions to consider for getting a good and reliable joint. It is observed that other than the weldability of the material, there are three additional factors that play vital roles during the USW process: part thickness, width, and length. It is reported that part thickness is an important factor that affects the weld strength significantly [10]. Thinner specimens have a better chance of being

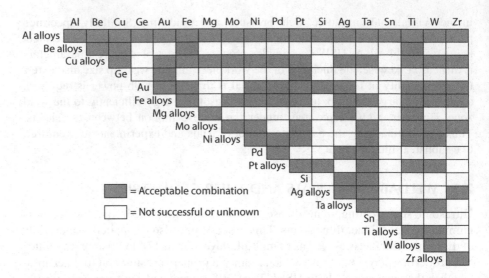

FIGURE 2.2 Ultrasonic metal weldability chart [11].

welded properly. However, as the thickness of the parts increases, a large sonotrode tip, more clamping pressure, and more electrical power are needed to achieve good weld joints.

On a high-power ultrasonic welder (typically 2-4 kW), it is possible to weld 6XXX and 5XXX aluminum alloy series metal sheets of 1-2 mm thickness. The specimen dimensions also play a major role in achieving superior welds. Thus, the dimensions should be carefully selected, because the alternating high-frequency vibration transmits through the weld joints, and the resonance varies with the increase in dimensions. The details of these three factors are reviewed next.

Specimen thickness: In most cases, the specimens are placed in an overlapped manner and parallel to the ultrasonic vibration [4]. In ultrasonic metal welding (USMW), the ultrasonic energy passes through the top workpiece to the weld spot with a loss of energy due to friction, resulting in the generation of internal heat. The vibrational amplitude of the process has to be adjusted to the thickness of the specimen. The relationship between the energy losses that occur during welding and the thickness of the part was given by Wodara [12] as

$$P_x = P_0.e^{\frac{-k.x}{20}} \tag{2.1}$$

where P_x and P_0 are the alternating acoustic pressures, k is the absorption coefficient, and x is the thickness.

This acoustic pressure is superimposed with the normal and shear force exerted by the sonotrode tip on the top specimen, and it gradually decays with the increase in part thickness. In experiments, it has been noticed that weld strength decreases exponentially with the increase in part thickness at a constant output power [5], as illustrated in Figure 2.3. Curve 1 and 2 represent the 80-μm and 20-μm-thick bottom sheets on which the USW experiments were carried out.

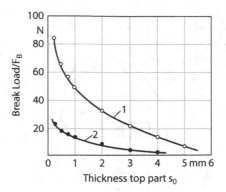

FIGURE 2.3 Variation of weld strength with the thickness of top specimen [5].

Specimen width: The *width* of the top specimen refers to the dimension parallel to the ultrasonic vibration. When the weld is made exactly at the center of the workpiece, and its dimension is half of the wavelength, a significant decrease in weld strength can be noticed. The wavelength of sound propagated through a material can be calculated from equation 2.2:

$$\lambda = \frac{c}{f} = \frac{\sqrt{\frac{E}{\rho}}}{f} \tag{2.2}$$

where c is the sound speed in a rod, E is Young's modulus, ρ is the density, and f is the frequency.

If the width of the specimen is equivalent to $\lambda/2$, the vibration occurs at the nodal position of the part. The results of an experiment on 0.5-mm-thick copper sheets revealed that the weld strength decreases with the increase in width due to the decrease in wavelength and increase in the frequency of the sound propagated in the material [5]. The results are presented in Figure 2.4. Curve 1 in Figure 2.4 represents

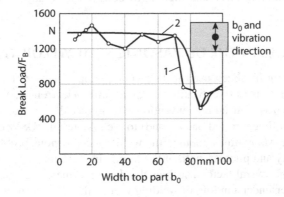

FIGURE 2.4 Impact of top part width on weld strength [5].

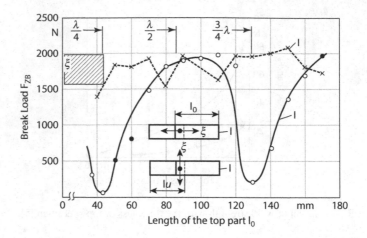

FIGURE 2.5   Relation between top part length and failure load [5].

the measured break loads of the ultrasonically welded joints, and curve 2 indicates the theoretical breaking loads with respect to width of the top part.

Specimen length: The length of the weld specimen also influences the soundness of the joint when its direction is parallel to the sonotrode vibration direction. When the experiments are performed at the edge of the specimens, it is observed that the critical length of the part was $(2n + 1)\frac{\lambda}{4}$ (n represents the number of modes and λ is the wavelength of the sound wave) when the weld strength noticeably decreased [13] (Figure 2.5, in which curves I and II represent theoretical and measured breaking loads of the ultrasonically welded joints with respect to the length of weld specimen, respectively.

At this length, the sonotrode deforms at the top part comprehensively without welding. If the length of the top part is $n\frac{\lambda}{2}$ then the top part is vibrating in resonance with the other components. In this case, a little deformation with very good welding can be achieved. However, if the length of the top specimen is perpendicular to the sonotrode vibration direction, the length does not have any effect on the weld strength.

2.3   KEY ISSUES IN THE ULTRASONIC SPOT WELDING PROCESS

Ultrasonic welding has been used in fabrication shops for more than 20 years, in manufacturing industries like aviation, medical, microelectronics, and many others due to various hurdles raised by conventional fusion welding processes. As it takes only a very short time (around one second) to weld materials, USW can be used for mass production. Many times, though, the welder faces various problems leading to poor weld quality and poor strength of the joints. In fact, the quality and success of a weld depend on several factors, such as the design of the ultrasonic welding components, a proper understanding of welding mechanics, selection of materials, and control of process parameters.

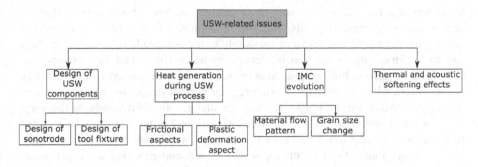

FIGURE 2.6 A schematic presentation of various issues in the USW process.

A conceptual diagram of the important scientific issues related to USW is provided in Figure 2.6. Four key components—ultrasonic system component design, heat generation, intermetallic compound (IMC) evolution, and acoustic softening—influence the weld strength. All four of these components affect the interface temperature, often resulting in a convoluted flow pattern and microstructural changes in the grain structure. Thus, there is a need for robust understanding of these components as we establish a precise relationship among the control factors (e.g., weld pressure, weld time, weld energy, vibration amplitude, etc.) and performance measures (e.g., tensile shear strength, temperature, etc.).

2.3.1 ULTRASONIC WELDING SYSTEM COMPONENT DESIGN

The discovery of piezoelectric crystals resulted in successful implementation in the ultrasonics field for various applications. In subsequent years, the intensity of the acoustic power has been increased to produce high-power acoustics, achieved through the development of different allied components [14, 15]. One of the important parts of the ultrasonic welding system is the acoustic horn/sonotrode. It primarily serves two functions: magnifying the amplitude of mechanical vibration required for welding and also acting as a tool which is directly in contact with the workpiece. The ultrasound waves are transmitted through the sonotrode and applied to the welding surface. Thus, any oxide layers or other contaminants on the faying surface are dispersed due to high-frequency vibration, and proper metal-to-metal contact takes place. By that time, the metals have been heated due to interatomic vibration and interfacial friction, which promotes deformation of the material. Many researchers are focused on the various horn design criteria and have performed numerous experiments on different horns to validate their views. In most cases, equations for the horn are derived to compute its resonant length and amplitude gain. Such an analysis is carried out by taking a non-uniform bar, which was excited in a longitudinal vibration mode. This study also demonstrated the effects of the various horn shapes, such as conical, exponential, and catenoidal. Out all of these, catenoidal horns showed the largest amplitude gain [16]. With the support of such studies, many researchers conducted ultrasonic machining experiments using different horns, such as the cylindrical, tapered, and exponential types. Modal and harmonic analysis of horns was also carried out by using finite element analysis. It was noticed that the proper sonotrode shape depends upon the fundamental factors like resonance frequency and the amplitude factor at the output of the sonotrode tip. Moreover, these sonotrode dimensions were also influenced by the stiffness and slenderness ratio [17]. Thus, for getting high amplitude at the sonotrode tip, a new type of horn design is considered, which is superior to other regular horn designs. Generally, resonance frequency depends on the length of the horn, and amplitude depends on the varying cross-sectional area. To manufacture the various horns, size must be reduced from larger diameters to a small horn tip diameter, which may be time-consuming. Thus, a novel folded-type sonotrode was designed and modeled for applications where reduction in dimensions and weight is the prime concern [18]. Furthermore, in order to obtain maximum amplitude magnification, and a higher material removal rate with safe working stress in ultrasonic machining of hard and brittle materials, a new type of horn design was suggested with the help of finite element analysis. The newly designed sonotrode was manufactured without the computer numerical controlled (CNC) machines, and its design was much simpler than that of the exponential sonotrode. It also provided higher amplitude magnification than traditionally designed sonotrodes [19]. A schematic representation of various traditionally designed horns is presented in Figure 2.7.

Various mode shapes, such as longitudinal and torsional mode shapes, for the ultrasonic exponential horn were analyzed by different researchers. When the plane wave propagation equation for the exponential horn is considered with a proper decay constant of the area, both mode shapes may be observed. This analytical calculation is also validated by conducting experiments [20]. Conical, stepped, exponential, gaussian, and catenoidal horns are the most-used horn shapes in many industries. Of all of these types, the stepped horn has the largest amplification ratio. The amplification factor for catenoidal horns is higher than for conical, exponential, and gaussian types. However, the mass of various concentrators/horns (such as conical, exponential, and catenoidal) determines their longitudinal mode of vibration [21]. To support

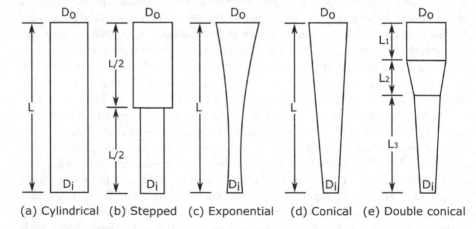

(a) Cylindrical   (b) Stepped   (c) Exponential   (d) Conical   (e) Double conical

FIGURE 2.7   Traditionally designed horn contours.

this study, a mathematical analysis with the horn dimensions was performed and showed that the maximum gain could best be achieved by a stepped horn as compared to other ones [22]. Meanwhile, it has also been shown in ultrasonic machining experiments that the exponential horn demonstrated the amplification value more than other horn profiles. However, this type of horn is difficult to manufacture [23]. For high-power ultrasonic applications, the complete design formulas for the conical steel section are derived for uncovering particle velocity, stress, mechanical impedance, amplification factor, and length. The formula is based on the assumptions that its length is half of its wavelength and that it operates at 20 kHz. Thus, an amplification of 4.61 can be gained [24]. The design and finite-element modeling of a stepped horn have also been performed for ultrasonic aided electro-discharge machining. In these studies, the resonant length of a horn was discovered by adjusting the length of the horn, and the effect of the radius of curvature on the natural frequency of the horn was also revealed. The study results show that with an increase in the radius of curvature, and a decrease in length, the natural frequency of a horn increases. Ultrasonic horns are used not only for cutting, drilling, or welding operations, but also could be used for embedding and encapsulating metals into joined specimens. Thus, the correct design is crucial in order to get a defect-free and good-quality joint [25, 26]. In contrast to conventionally designed horns, the modal vibration analysis of rectangular horns revealed that the length of the horn could be set to half of the wavelength with a longitudinal vibration of 20 kHz. The uniform amplitude was attained with the reduced Poisson's ratio at the weld tip surface by adding a number of slits. The previous study concluded that the frequency of longitudinal mode depends on slit width and thickness of bridge [27].

Although study of the effect of various horn designs is popular in cutting and drilling processes, it is less explored in the ultrasonic welding process. An attempt was made to investigate the various horn profiles numerically for welding of thermoplastic components. It was found that the stepped horn utilized 75% of the normal energy as compared to 85% for the Bezier horn, and the Bezier horn produced more interface temperature than the stepped one. The study also showed that Bezier-profiled horns developed less stress than the stepped horn, due to its smooth curvature at the nodal position. Thus, it could be useful for welding of thicker high-density polyethylene (HDPE) specimens [28]. The comparison graph of various horn design effects during the ultrasonic plastic welding of HDPE specimens is presented in Figure 2.8.

Other numerical studies [29–31] focused on different horn profiles, such as Bezier and nontraditional B-spline curve, with an open uniform knot vector (OUNBS). These improved horns had higher displacement amplification than the catenoidal horn and lower stress concentration than the stepped horn. These studies explore the resonant frequencies and amplitude magnifications of the horn, which depend upon its shape, length, material, and density. Ultimately, these amplitude magnifications at the sonotrode tips determine the effectiveness of the bond.

2.3.2 HEAT GENERATION DURING THE USW PROCESS

The analysis of heat generation at the horn-top workpiece and the top workpiece-bottom workpiece interfaces during the ultrasonic welding of similar materials is

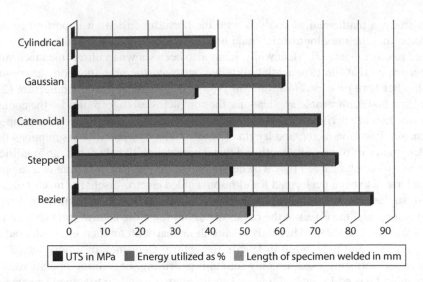

FIGURE 2.8 Effect of various horn designs on ultrasonic plastic welding of HDPE specimens [28].

somewhat more straightforward than for the ultrasonic welding of dissimilar materials, because there are fewer differences in the physical, thermal, and mechanical properties of the work materials. These properties severely affect heat generation at the interfaces, temperature distribution, and material flow in the workpieces. Hence, in ultrasonic welding of dissimilar materials, the stacking position of the sheets becomes crucial. Typically, the material that is less hard is placed on the sonotrode side because it will be more plastically deformed than the lower sheet in a similar working environment. The existing literature on ultrasonic welding of dissimilar Al to steel sheets suggests that the Al should be placed on the sonotrode side; the sonotrode tip plunges more into this material. Moreover, a considerable lap shear strength can be attained due to the generation of high interface temperature, leading to severe plastic deformation and drastically reduced material yield strength. Furthermore, as the Al sheet softens more, the mutual rubbing action at the interface increases, resulting in a rapid increment of micro welds at the interface, and these are saturated with an increase in weld time. The heat accumulation and the shear deformation mostly occur at the Al side because of the lower thermal conductivity of steel. Thus, a suitable strategy must be devised to prevent any type of crack formation due to severe reduction of material yield strength and excessive plastic deformation.

The next issue usually faced in ultrasonic welding of dissimilar materials is the propagation of heat in the weld materials. In similar-material welding, the thermal profile is symmetric around the weld center. However, in dissimilar welding, the thermal profile is asymmetric due to substantial variation in the thermal conductivity of the materials. The presence of different zones for a given set of working conditions is schematically illustrated in Figure 2.9.

The three distinct types of zones can be noticed in this figure. The weld zone (WZ) is the area where severe plastic deformation occurred due to the friction between

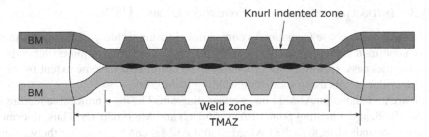

FIGURE 2.9   Schematic illustration of different weld zones in USW process.

the weld sheets. As a result, the recrystallization of grains took place due to a sudden increase in the interface temperature. The zone adjacent to the WZ is the thermo-mechanically affected zone (TMAZ). This is the region up to which the heat propagates. Adjacent to this zone, the base metal (BM) starts. In this zone, the microstructural and mechanical properties are unaffected by the welding process. The most noticeable feature is the growth of the WZ and TMAZ with respect to increases in weld energy values. At lower weld energy, there is no distinct increase in the WZ and TMAZ regions, due to the inadequate thermal softening at the weld interface.

Figure 2.10 reveals that the TMAZ size (taking half of its size in this case) sharply increases with weld energy. However, the WZ size initially shows a sluggish type of increment and finally becomes constant with the rise in weld energy. Thus, at higher weld energy, there is a sharp increase in the interface temperature near the material with lower thermal conductivity. It results in the achievement of a high softening temperature, which ultimately leads to substantial variations in flow characteristics for two dissimilar metals. It might cause sticking of the weld materials to the sonotrode tip or anvil and formation of cracks around the periphery of the WZ.

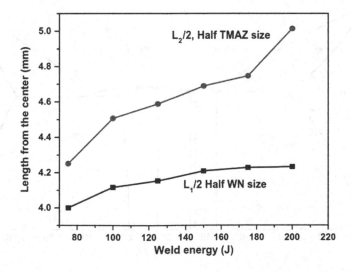

FIGURE 2.10   Variation of TMAZ and weld nugget size with weld energy [32].

2.3.3 Intermetallic Compound Evolution During USW

The intermetallic phase formation in conventional fusion welding of dissimilar metals is a common phenomenon that severely affects the integrity of joints, depending upon its thickness. However, this problem can be eliminated to some extent by joining dissimilar metals using solid-state welding techniques.

Figure 2.11 shows an Al-Mg binary phase diagram. The most noticeable feature is the nearly identical melting point of Al (~660°C) and Mg (~650°C). Thus, intermetallic compounds (IMCs) such as $Al_{12}Mg_{17}$ and Al_3Mg_2 can form during the welding process. The microhardness of these IMCs is on the scale of 152-221 HV, in comparison to the 25-60 HV of base metals [33]. This reveals the brittleness of the IMCs at the interface, which severely degrades the joint strength and forms the fracture. Therefore, it is very challenging to attain a robust and non-brittle joint between Al and Mg alloys. The critical approach for enhancing the weld strength results is to reduce the formation of brittle IMCs as much as possible during the USW process. There are three ways to minimize the negative effect of IMCs: (1) shorter weld time, (2) lower weld interface temperature, and (3) use of a third material as an interlayer that probably reacts with both Al and Mg based upon their respective binary phase diagrams.

Likewise, Figure 2.12 depicts the Al-Fe binary phase diagram. This figure highlights three major facts: (1) a wide difference in the Al and Fe melting points, (2) existence of a variety of IMCs, and (3) negligible solubility of Al in the Fe lattice at operating temperature. IMCs such as $FeAl_3$, Fe_2Al_5, $FeAl$, Fe_3Al, and $FeAl_2$ can be formed at different temperatures during the USW process. Thus, the mechanical

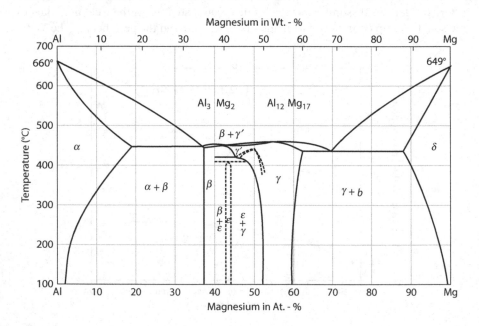

FIGURE 2.11 Al-Mg binary phase diagram [34].

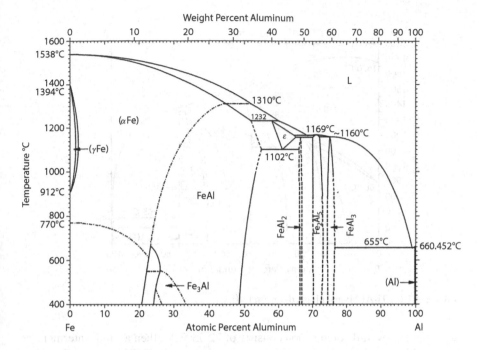

FIGURE 2.12   Fe-Al binary phase diagram [37].

strength of the joint is critically affected by the thickness, size, and type of inter-metallic phases formed in the weld zone. Some studies revealed that the growth of brittle IMCs during USW of Al-Fe metals could be inhibited by means of an inter-layer or coating at the weld interface [35, 36]. Therefore, it is necessary to understand the factors affecting the formation of IMCs qualitatively.

It is observed from these studies [35, 36] that the formation of intermetallic phases occurred at a specific temperature during the USW of dissimilar metals. In this context, the effective Gibbs free energy (ΔG_p^e) concept can be utilized for effec-tive prediction of a most suitable intermetallic phase at a particular temperature. The expression of ΔG_p^e for the intermetallic phase p is expressed as

$$\Delta G_p^e = \Delta G_p \times \frac{C_e}{C_p} \tag{2.3}$$

where ΔG_p is the Gibbs free energy for the formation of phase p, C_e is the limiting element concentration at the interface, and C_p is the limiting element concentration at the IMC zone. The intermetallic phase, which has the most negative ΔG_p^e value, is favored at a particular temperature. Here, an Al-Ti binary system is used for the discussion of the Gibbs free energy concept on the intermetallic phases during the USW process. The Al-Ti binary phase diagram is shown in Figure 2.13.

This diagram presents the possible formation of four intermetallic compounds (Ti_3Al, $TiAl_2$, $TiAl_3$, and $TiAl$) at 665°C. The formation of these compounds also depends upon the atomic fractions of Al and Ti. If the Al and Ti are mixed in such

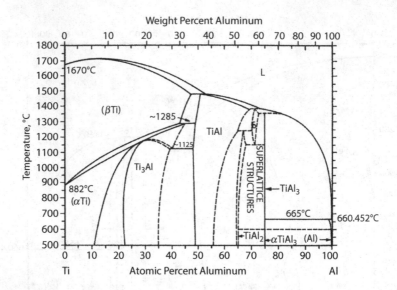

FIGURE 2.13 Ti-Al binary phase diagram [38].

a way that the overall composition consists of 72-75% Al, then a $TiAl_3$ intermetallic phase will be produced. As the welding temperature in the USW process is less than the melting point of these parent metals, the $TiAl_3$ phase is dominant [39, 40] over other phases. Table 2.1 presents the Gibbs free energy calculation for three compositions. It may be observed from this table that the $TiAl_3$ intermetallic phase has the most negative value, and this phase will be favored at 400°C (a peak temperature reported in previous literature [41]).

2.3.4 ACOUSTIC SOFTENING EFFECTS ON WELD STRENGTH

Acoustic softening is the second cause of material softening during the USW process. It was first identified through experiments in which zinc monocrystals were exposed to ultrasonic energy of 800 kHz [43]. It was observed that at a sufficient level of ultrasonic energy, the yield stress of the material was reduced. Likewise, several quasi-static compression experiments on aluminum alloys showed a similar type of reduction in the

TABLE 2.1

Effective Gibbs Free Energy Calculation for Various Phases of Al-Ti Binary System at 400°C [42]

Phase	ΔG_p^e (J/mol)	ΔG_p^e (400°C) (kJ/mol)
Ti_3Al	−29633.6+6.70801*T	−26.95
TiAl	−37445.1+16.79376*T	−30.72
$TiAl_3$	−40349.6+10.36525*T	−36.20

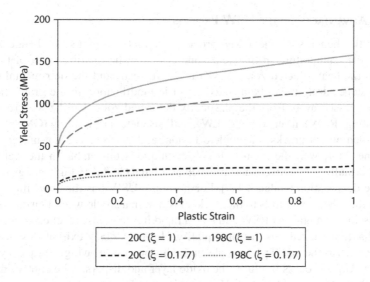

FIGURE 2.14 Comparison of stress-strain results for AA1100 aluminum alloy considering thermal and acoustic softening parameter [48].

yield stress value in the presence of ultrasonic energy over a broad range of frequencies [44, 45]. Acoustic softening of a material depends on several factors, including acoustic impedance, melting point, Young's modulus, and hardenability. However, acoustic softening is more active on materials with high acoustic impedance and Young's modulus values with a low melting point. The finite element model was also utilized to investigate the plastic deformation and stress fields during USW regarding the acoustic softening phenomenon. Experiments revealed that the plastic deformation and the temperature generation during welding were due not merely to the thermal softening phenomenon, but that acoustic softening also played a key role in getting an accurate prediction of interface temperature [46]. In the ultrasonic consolidation (UC) process, thermo-mechanical modeling along with the acoustic softening parameter are employed to explore the severe plastic deformation in the bonding region. A correlation was also found between plastic strain from this model and experimentally bonded areas [47]. Figure 2.14 demonstrates the relative contributions of thermal and acoustic softening effects on the stress-strain profile of AA 1100 aluminum material during the UC process. It can be inferred from the figure that the results without acoustic softening ($\xi = 1$) severely underpredict the stress values as compared to the results with acoustic softening ($\xi = 0.177$). Yield stress results are reduced 82.3% due to acoustic softening. Thus, the role of acoustic softening in the USW process cannot be ignored. Hence, this critical approach can be considered to model the interface temperature profile accurately, and ultimately the exact reasons for plastic deformation can be revealed.

2.4 PROS AND CONS OF THE USW PROCESS

Though unlike other welding processes, the USW process also has its own advantages and limitations. These details are discussed in the following sections.

2.4.1 ADVANTAGES OF THE USW PROCESS

Most of the benefits of the USW process originate from its solid-state nature. Because the temperature in the weld zone is below the melting point of the base material, the heat-affected zone (HAZ) is diminished, and the detrimental effects of stress concentration can be avoided. A little modification in the grain structure can be witnessed away from the plastically deformed zone. Unlike with resistance spot welding (RSW), laser welding (LW), and gas metal arc welding (GMAW), weld defects such as hot cracks, incomplete fusion, and porosity are not present in the weld zone. Moreover, the popular RSW technique is unsuitable in the welding of highly thermally conductive materials such as Al and Cu due to high energy cost and electrode degradation tendency [49]. Furthermore, USW has the capability to weld multiple thin sheets and foils to the thick substrate in a single weld. During the joining of dissimilar metals by RSW and LW, a rapid intermetallic reaction occurs in the liquid phase, and it reduces the joint quality. However, USW exhibits less tendency to form IMCs, as the liquid phase is absent in this process. The high-frequency vibration of the USW process fractures the oxide layer and disperses the contaminants to the periphery of the weld nugget. Hence, this process may not require any particular cleaning process at the weld zone, though proper attention must be paid to the steadiness of surface conditions. There is no requirement of cooling time or setting time in the USW process, and it takes only a fraction of a second (typically 0.2 to 0.5 sec) to complete the weld cycle. This feature of USW makes it ideal for automation. No shielding gases or filler metals are needed in this process. However, a thin interlayer sheet can be employed to join hardened metals effectively. Meanwhile, the RSW method uses quite a bit more energy (50-100 kJ) than USW, and friction stir spot welding (FSSW) involves both high energy (3-6 kJ) and longer weld period (~ 2.5 sec) as compared to the USW process (0.6-1.5 kJ) [50, 51]. With advancements in lightweight vehicles, the use of Al and Mg alloys become inevitable in the automotive industries. The large automotive makers such as Ford Motor Company have investigated the feasibility of USW and compared the relative cost per joint of the USW process with RSW, adhesive bonding, GMAW, and self-piercing rivet (SPR) processes during the assembly of body parts. It was shown that the USW process outperforms other methods, as it is cost-efficient, reliable, and environmental friendly [52].

2.4.2 LIMITATIONS OF THE USW PROCESS

A sufficient amount of friction at the weld interface can happen only if the top part moves perceptibly with the ultrasonic horn relative to the bottom part. This movement is only achieved through a lap joint. It is not possible to apply currently developed USW systems to produce butt, tee, or corner joints for metal sheets. A good weld quality can be achieved for up to 2 mm thickness of aluminum alloys, and 0.4 mm thickness of harder alloys like titanium and steel alloys. The ultrasonic welding tip penetrates the softer top workpiece during the welding process. This creates deformations and reduces the yield strength of the part. Thus, special attention should be given to horn as well as anvil design, to find those that will provide less distortion or marking. Irritating noise during the process is another issue faced by the welder. Although the ultrasonic system works at a frequency of 20 kHz, subharmonic vibrations may be induced in the weld materials. Thus, an acoustic enclosure is essential to shield workers from the radiated sound waves.

REFERENCES

1. Weisman C. Welding handbook, vol 5, 7th edition, American Welding Society, Miami FL, 1997.
2. Harthoorn JL. Ultrasonic metal welding (Thesis). Eindhoven, Tech Hogeschool, Dr Tech Wet Diss. 1978 150 P 1978.
3. Janaki Ram GD, Robinson C, Yang Y, Stucker BE. Use of ultrasonic consolidation for fabrication of multi-material structures. Rapid Prototyp J 2007;13:226–35.
4. Bloss MC. Ultrasonic Metal Welding: The Weldability of Stainless Steel, Titanium, and Nickel-Based Superalloys. The Ohio State University. 2008.
5. De Vries E. Mechanics and mechanisms of ultrasonic metal welding. The Ohio State University. 2004.
6. Derks PLLM. The design of ultrasonic resonators with wide output cross-sections. Technische Hogeschool Eindhoven. 1984.
7. Graff KF, Devine JF, Kelto J, Zhou NY. Ultrasonic welding of metals. vol. 3. United Kingdom: Elsevier Ltd.; 2007. doi:10.1016/B978-1-78242-028-6.00011-9.
8. Bloss M, Graff K. Ultrasonic metal welding of advanced alloys: the weldability of stainless steel, titanium, and nickel-based superalloys. Trends Weld Res Proc 8th Int Conf 2009, pp. 348–53.
9. Patel VK, Bhole SD, Chen DL, Ni DR, Xiao BL, Ma ZY. Solid-state ultrasonic spot welding of SiCp/2009Al composite sheets. Mater Des 2015;65:489–95.
10. Ahmed N. New developments in advanced welding. 1st ed. England: CRC Press; 2005.
11. Graff KF, Devine JF, Keltos J, Zhou NY. Ultrasonic welding of metals. In: O'Brien AO, ed. Weld. Handb. 9th ed., Miami, FL: American Welding Society; 2007, pp. 263–303.
12. Wodara J. Relation between component geometry and the quality of ultrasonically welded metallic joints. Wissenschaftliche zeitschrift der tech univ otto von. Guericke Magdebg 1989;33:64–68.
13. Okada M, Shin S, Miyagi M, Matsuda H. Joint mechanism of ultrasonic welding. Trans Japan Inst Met 1963;4:250–55.
14. Lucas M, Cardoni A, McCulloch E, Hunter G, MacBeath A. Applications of power ultrasonics in engineering. Appl Mech Mater 2008;13:11–20.
15. Shuyu L. Sandwiched piezoelectric ultrasonic transducers of longitudinal-torsional compound vibrational modes. IEEE Trans Ultrason Ferroelectr Freq Control 1997;44:1189–97.
16. Merkulov LG. Design of ultrasonic concentrations. Sov Physics-Acoustics 1957;4:246–55.
17. Nad M. Ultrasonic horn design for ultrasonic machining technologies. Appl Comput Mech 2010;4:79–88.
18. Sherrit S, Askins SA, Gradziol M, Dolgin BP, Bao X, Chang Z, et al. Novel horn designs for ultrasonic/sonic cleaning, welding, soldering, cutting, and drilling. SPIE's 9th Annu Int Symp Smart Struct Mater 2002, pp. 353–60.
19. Amin SG, Ahmed MHM, Youssef HA. Computer-aided design of acoustic horns for ultrasonic machining using finite-element analysis. J Mater Process Technol 1995;55:254–60. doi: 10.1016/0924-0136(95)02015-2.
20. Lin S. Study on the longitudinal-torsional composite mode exponential ultrasonic horns. Ultrasonics 1996;34:757–62.

21. Amza G, Drimer D. The design and construction of solid concentrators for ultrasonic energy. Ultrasonics 1976;14:223–6.
22. Belford JD. The stepped horn-Technical Publication TP-214. 2011.
23. Satyanarayana A, Reddy BGK. Design of velocity transformers for ultrasonic machining. Electr India 1984;24:11–20.
24. Ensminger D. Solid cone in longitudinal half-wave resonance. J Acoust Soc Am 1960;32:194–6.
25. Alexandru N, Niculae S, Marinescu I. Study on ultrasonic stepped Horn geometry design and FEM simulation. Nonconv Technol Rev 2011, pp. 25–30.
26. Shu K, Hsiang W, Chen CC. The design of acoustic horns for ultrasonic insertion. J Chinese Soc Mech Eng 2010;4:338–42.
27. Adachi K, Ueha S, Mori E. Modal vibration analysis of ultrasonic plastic welding tools using the finite element method. Proc Ultrason Int 1986, pp. 727–32.
28. Roopa Rani M, Rudramoorthy R. Computational modeling and experimental studies of the dynamic performance of ultrasonic horn profiles used in plastic welding. Ultrasonics 2013;53:763–72. doi: 10.1016/j.ultras.2012.11.003.
29. da Silva JB, Franceschetti NN, Adamowski JC. Numerical analysis of a High power piezoelectric transducer used in the cutting and welding of thermoplastic textiles. ABCM Symp Ser Mechatronics 2006;2:142–9.
30. Wang D, Chuang W-Y, Hsu K, Pham H-T. Design of a Bézier-profile horn for high displacement amplification. Ultrasonics 2011;51:148–56. doi: 10.1016/j.ultras. 2010.07.004.
31. Nguyen H-T, Nguyen H-D, Uan J-Y, Wang D-A. A nonrational B-spline profiled horn with high displacement amplification for ultrasonic welding. Ultrasonics 2014;54:2063– 71. doi: 10.1016/j.ultras.2014.07.003.
32. Shakil M, Tariq NH, Ahmad M, Choudhary MA, Akhter JI, Babu SS. Effect of ultrasonic welding parameters on microstructure and mechanical properties of dissimilar joints. Mater Des 2014;55:263–73.
33. Liu L, Ren D, Liu F. A review of dissimilar welding techniques for magnesium alloys to aluminum alloys. Materials (Basel) 2014;7:3735–57.
34. Murray JL. The Al-Mg (aluminum-magnesium) system. Bull Alloy Phase Diagrams 1982;3:60–74. doi: 10.1007/BF02873413.
35. Chen S, Huang J, Ma K, Zhang H, Zhao X. Influence of a Ni-foil interlayer on Fe/Al dissimilar joint by laser penetration welding. Mater Lett 2012;79:296–9.
36. Haddadi F, Strong D, Prangnell PB. Effect of zinc coatings on joint properties and interfacial reactions in aluminum to steel ultrasonic spot welding. JOM 2012;64:407–13.
37. Haidara F, Record MC, Duployer B, Mangelinck D. Phase formation in Al-Fe thin film systems. Intermetallics 2012;23:143–7. doi: 10.1016/j.intermet.2011.11.017.
38. Das S. The Al-O-Ti (aluminum-oxygen-titanium) system. J Phase Equilibria 2003;23:525–36. doi: 10.1361/105497102770331271.
39. Zhang LZ, Wang DN, Wang BY, Yu RS, Wei L. Annealing studies of Ti/Al multilayer film by slow positron beam. Appl Surf Sci 2007;253:7309–12.
40. Tardy J, Tu KN. Solute effect of Cu on interdiffusion in Al 3 Ti compound films. Phys Rev B 1985;32:2070.
41. Thuillard M, Tran LT, Nicolet M-A. Al3Ti formation by diffusion of aluminum through titanium. Thin Solid Films 1988;166:21–8.
42. Kattner UR, Lin JC, Chang YA. Thermodynamic assessment and calculation of the Ti-Al system. Metall Trans A, Phys Metall Mater Sci 1992;23 A:2081–90. doi: 10.1007/ BF02646001.
43. Blaha F, Langenecker B. Tensile deformation of zinc crystal under ultrasonic vibration. Naturwissenschaften 1955;42:0.

44. Kirchner HOK, Kromp WK, Prinz FB, Trimmel P. Plastic deformation under simultaneous cyclic and unidirectional loading at low and ultrasonic frequencies. Mater Sci Eng 1985;68:197–206.
45. Izumi O, Oyama K, Suzuki Y. Effects of superimposed ultrasonic vibration on compressive deformation of metals. Trans Japan Inst Met 1966;7:162–7.
46. Doumanidis C, Gao Y. Mechanical modeling of ultrasonic welding. Weld J 2004; 83:140–6.
47. Zhang C, Li L. A friction-based finite element analysis of ultrasonic consolidation. Weld J 2008;87:187.
48. Kelly GS, Advani SG, Gillespie JW. Bogetti T a. A model to characterize acoustic softening during ultrasonic consolidation. J Mater Process Technol 2013;213:1835–45. doi: 10.1016/j.jmatprotec.2013.05.008.
49. Peng J, Fukumoto S, Brown L, Zhou N. Image analysis of electrode degradation in resistance spot welding of aluminium. Sci Technol Weld Join 2004;9:331–6.
50. Bakavos D, Prangnell PB. Effect of reduced or zero pin length and anvil insulation on friction stir spot welding thin gauge 6111 automotive sheet. Sci Technol Weld Join 2009;14:443–56.
51. Su P, Gerlich A, North TH, Bendzsak GJ. Energy utilisation and generation during friction stir spot welding. Sci Technol Weld Join 2006;11:163–9.
52. Hetrik E. Ultrasonic spot welding—a new tool for aluminum joining. Weld J 2005; 84:26–30.

3 Tool and Fixture Design for Ultrasonic Spot Welding of Dissimilar Metal Sheets

Ultrasonic spot welding (USW) is an emerging joining technology that offers potential solutions in many industrial applications. The essential components of this equipment are the horn, booster, and anvil/fixture, through which the necessary pressure and high-frequency vibration are applied to create a solid-state weld joint. A good weld can be produced on the contact surfaces through a suitable scrubbing motion between the sheets, and this can be possible when the welding system components are correctly designed. The vibration amplitude at the output end of the transducer is very small (in the range of 10–30 μm), and welding does not take place in this amplitude range. Thus, the majority of USW systems employ a booster, which increases the vibration amplitude according to its design shape.

The booster also provides support for the mounting of the welding stack. The booster is made up of materials such as titanium, monel, and stainless steel, which have excellent acoustic properties and high strength. Another critical component of this welding system is the acoustic horn. It serves mainly two functions: it magnifies the amplitude of vibration obtained from the booster, and it also acts as a tool that is directly in contact with the workpiece. The common practice is to mount both components (booster and horn) parallel to the direction of vibration through which the ultrasound waves are transmitted and applied to the welding surface in a transverse mode. The objective of this chapter is to provide basic knowledge about the perfect design and manufacturing of these components to generate high ultrasonic energy to produce a sound weld. Proper equipment is crucial, as an incorrectly designed horn or booster may damage the whole system because of the mismatching of mechanical impedance. So far, the empirical method has been used to find out the optimum length and other dimensions of these components, but it is a very lengthy, time-consuming process, and may also not yield accurate results for the intricately shaped horn and booster. Therefore, finite element analysis with proper boundary conditions is preferred to analyze the complex geometries of horn and booster effectively.

3.1 THEORETICAL ASPECTS OF TOOL AND FIXTURE DESIGN

Most of the high-intensity and high-power applications of ultrasonics in manufacturing industries need a horn and a booster to provide a higher amplitude of vibration at the output (load) end. These two components act as the tools to concentrate

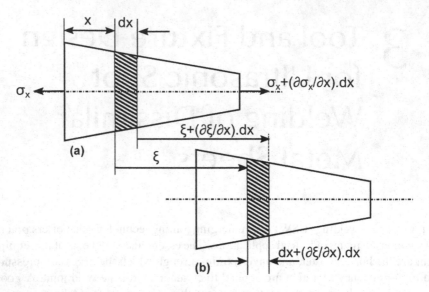

FIGURE 3.1   (a) Tapered, axisymmetric and slender bar; (b) changes in stress distribution after plane wave propagates through it.

the acoustic waves in a directional radiation pattern. The analysis of the pattern of waves through both of these pieces is very complicated, and specific assumptions are required for deriving the design equations even for a solid medium. The preferred mode of ultrasonic welding is the longitudinal mode, which can be analyzed by considering a free-free vibration in a non-uniform bar [1], as shown in Figure 3.1. The assumptions are:

i. The wave is propagated in the rod along the axial direction.
ii. The walls of these components are so rigid that wave propagation along the lateral direction is neglected.
iii. The amplitudes of acoustic pressure remain within the limit of compressibility characteristics for the solid. Therefore, the second-order differential terms can be neglected.
iv. The bar is tapered in such a way that the presumed plane waves can retain contact with its walls.

Let's take an infinitesimally small element on this non-uniform bar of length dx. If ξ is the displacement at x, then at the $x + dx$ position, it will be $\xi + (\partial\xi/\partial x)\, dx$. It is then evident that the element dx in the new position has changed by an amount $(\partial\xi/\partial x)\, dx$. Thus, the strain developed in the strip is:

$$Strain = \frac{\left(\xi + \frac{\partial\xi}{\partial x}dx\right) - \xi}{dx} \tag{3.1}$$

$$Strain = \frac{\partial \xi}{\partial x} \tag{3.2}$$

From Hooke's law of stress-strain relationship,

$$\sigma_x = E.\frac{\partial \xi}{\partial x} \tag{3.3}$$

Differentiating Eq. (3.3) with respect to x,

$$\frac{\partial \sigma_x}{\partial x} = E.\frac{\partial^2 \xi}{\partial x^2} \tag{3.4}$$

Due to the vibration, an accelerating force is exerted on the elementary strip, and according to Newton's law of motion, it can be represented as

$$F_a = m.a \tag{3.5}$$

$$F_a = \rho \times \vartheta \times \frac{\partial^2 \xi}{\partial t^2} \tag{3.6}$$

$$F_a = \rho \times S \times \frac{\partial^2 \xi}{\partial t^2}.dx \, (\text{where thickness is unity}) \tag{3.7}$$

Due to the non-uniform nature of the rod, the constraining force acting on the strip is

$$F_c = F_{final} - F_{Initial} \tag{3.8}$$

$$F_c = (\sigma \times S)_{final} - (\sigma \times S)_{initial} \tag{3.9}$$

$$F_c = \left[\left(\sigma_x + \frac{\partial \sigma_x}{\partial x}.dx \right) \times \left(S + \frac{\partial S}{\partial x}.dx \right) \right] - \sigma_x \times S \tag{3.10}$$

$$F_c = S \times \frac{\partial \sigma_x}{\partial x}.dx + \sigma_x \times \frac{\partial S}{\partial x}.dx + \frac{\partial \sigma_x}{\partial x}.dx \times \frac{\partial S}{\partial x}.dx \tag{3.11}$$

Neglecting the $(dx)^2$ term from Eqs. (3.10) and (3.11),

$$F_c = S \times \frac{\partial \sigma_x}{\partial x}.dx + \sigma_x \times \frac{\partial S}{\partial x}.dx \tag{3.12}$$

To satisfy the equilibrium of that small element, constraining force should be equal to the accelerating force, so by comparing Eqs. (3.7) and (3.12)

$$F_c = F_a \tag{3.13}$$

$$S \times \frac{\partial \sigma_x}{\partial x}.dx + \sigma_x \times \frac{\partial S}{\partial x}.dx = \rho \times S \times \frac{\partial^2 \xi}{\partial t^2}.dx \qquad (3.14)$$

$$S \times \frac{\partial \sigma_x}{\partial x} + \sigma_x \times \frac{\partial S}{\partial x} = \rho \times S \times \frac{\partial^2 \xi}{\partial t^2} \qquad (3.15)$$

Putting the values of Eqs. (3.3) and (3.4) in Eq. (3.15), we get

$$S.E.\frac{\partial^2 \xi}{\partial x^2} + E.\frac{\partial \xi}{\partial x} \cdot \frac{\partial S}{\partial x} = \rho \times S \times \frac{\partial^2 \xi}{\partial t^2} \qquad (3.16)$$

Dividing $\rho.S$ on both sides of Eq. (3.16)

$$\frac{E}{\rho} \cdot \frac{\partial^2 \xi}{\partial x^2} + \frac{E}{\rho.S} \cdot \frac{\partial \xi}{\partial x} \cdot \frac{\partial S}{\partial x} = \frac{\partial^2 \xi}{\partial t^2} \qquad (3.17)$$

Putting $V_s = \sqrt{\frac{E}{\rho}}$ in Eq. (3.17)

$$\frac{\partial^2 \xi}{\partial x^2} + \frac{1}{S} \cdot \frac{\partial \xi}{\partial x} \cdot \frac{\partial S}{\partial x} = \frac{\rho}{E} \cdot \frac{\partial^2 \xi}{\partial t^2} \qquad (3.18)$$

$$\frac{1}{V_s^2} \cdot \frac{\partial^2 \xi}{\partial t^2} + \frac{1}{S} \cdot \frac{\partial \xi}{\partial x} \cdot \frac{\partial S}{\partial x} - \frac{\partial^2 \xi}{\partial x^2} = 0 \qquad (3.19)$$

As both elements undergo harmonic motion, the displacement equation shown in Eq. (3.19) can be written as

$$\xi = A.\sin(\omega t) + B.\cos(\omega t) \qquad (3.20)$$

$$\frac{\partial \xi}{\partial t} = A.\omega.\cos(\omega t) - B.\omega.\sin(\omega t) \qquad (3.21)$$

$$\frac{\partial^2 \xi}{\partial t^2} = -\omega^2[A.\sin(\omega t) + B.\cos(\omega t)] \qquad (3.22)$$

$$\frac{\partial^2 \xi}{\partial t^2} = -\omega^2.\xi \qquad (3.23)$$

Putting the value of Eq. (3.23) in Eq. (3.19), we get

$$\frac{1}{V_s^2} \cdot (-\omega^2.\xi) - \frac{1}{S} \cdot \frac{\partial \xi}{\partial x} \cdot \frac{\partial S}{\partial x} - \frac{\partial^2 \xi}{\partial x^2} = 0 \qquad (3.24)$$

$$\frac{\partial^2 \xi}{\partial x^2} + \frac{1}{S} \cdot \frac{\partial \xi}{\partial x} \cdot \frac{\partial S}{\partial x} + \frac{\omega^2}{V_s^2}.\xi = 0 \qquad (3.25)$$

Eq. (3.25) is used to find out the resonant lengths of both axisymmetric horn and booster. Those can be determined by:

$$V_S = \lambda.f \tag{3.26}$$

Usually, the lengths of these parts are the half of the wavelength passing through them. The equations used to find out the resonant length of the different horn and booster profiles were obtained by Merkulov [2]. The following equations for calculating resonant length and stress can be achieved for different types of profiles.

i. *For stepped profile type*

$$L = \frac{\lambda}{2} = \frac{V_S}{2f} \quad \text{and} \tag{3.27}$$

$$Stress(S) = \frac{E\omega\xi_m}{V_S}\sin\left(\frac{\omega x}{V_S}\right)\sin(\omega t) \tag{3.28}$$

Stress is maximum at $x = \frac{l}{2} = \frac{\lambda}{4}$ position.

ii. *For exponential tapered type*

$$L = \frac{V_S}{2f}\sqrt{1+\left[\frac{\ln\left(\frac{s_2}{s_1}\right)}{2\pi}\right]^2} \tag{3.29}$$

$$Stress(S) = \frac{E}{\omega}V_0\left(\frac{\omega}{V_S'}+\frac{\gamma^2 V_S'}{4\omega}\right)e^{\frac{\gamma x}{2}}.\sin\left(\frac{\omega x}{V_S'}\right) \tag{3.30}$$

Maximum stress occurs at $\frac{dS}{dx}=0$ and $\tan\frac{\omega x}{V_S'}=\frac{-2\omega}{\gamma V_S'}$.

iii. *For catenoidal type*

$$L = \frac{V_S}{2\pi f}\sqrt{\pi^2+\left[\cosh^{-1}\frac{r_2}{r_1}\right]^2} \quad \text{and} \tag{3.31}$$

$$Stress(S) = \frac{EV_1\cos K'x}{\omega a\cosh\left(\frac{Kx}{a}\right)}\left(\tanh\frac{Kx}{a}+aK'\tan K'x\right) \tag{3.32}$$

At the point of maximum stress, $\frac{dS}{dx}=0$, from which $\frac{1}{S}\frac{dS}{dx}=\frac{2x}{x^2+b^2}$.

The one-dimensional propagation of the longitudinal wave in the bar can be analyzed by Eq. (3.18). Thus, it can be expressed as

$$\frac{\partial^2\xi}{\partial x^2}=\frac{\rho}{E}\cdot\frac{\partial^2\xi}{\partial t^2} \tag{3.33}$$

The solution to the above second-order differential equation depends on two variables.

$$\xi(x,t) = \xi(x).\xi(t) \tag{3.34}$$

The left-hand side of Eq. (3.33) is independent of x, whereas the right-hand side is independent of t. Thus, the simplified equation can be derived further:

$$\frac{\partial^2 \xi}{\partial x^2} + \left(\frac{\omega}{V_s}\right)^2 .\xi = 0 \tag{3.35}$$

$$\frac{\partial^2 \xi}{\partial t^2} + \omega^2 .\xi = 0 \tag{3.36}$$

According to Eq. (3.34), the general solution can be presented as

$$\xi(x) = A.\sin\left(\frac{\omega}{V_S}\right)x + B.\cos\left(\frac{\omega}{V_S}\right)x \tag{3.37}$$

$$\xi(t) = C.\sin(\omega t) + D.\cos(\omega t) \tag{3.38}$$

where A, B, C, and D are constants. So, the overall solution becomes

$$\xi(x,t) = \left[A.\sin\left(\frac{\omega}{V_S}\right)x + B.\cos\left(\frac{\omega}{V_S}\right)x\right].\left[C.\sin(\omega t) + D.\cos(\omega t)\right] \tag{3.39}$$

The natural frequency of both components can be found from Eq. (3.39) by putting the boundary conditions for the free-free bar. It can be written as

$$f_n = \frac{n_m}{2L}\sqrt{\frac{E}{\rho}} \tag{3.40}$$

The performance measure of the horn is the amplification ratio/gain and is given by

$$\frac{\xi_1}{\xi_2} = \frac{V_1}{V_2} = \frac{S_2}{S_1} = \left(\frac{D_2}{D_1}\right)^2 \tag{3.41}$$

From all the design profiles of horn and booster, the stepped profile offers a high amplification ratio as compared to exponential and conical profiles [3]. The finite element method is a suitable route by which to design and model these ultrasonic components more accurately and efficiently.

3.2 APPLICATION OF NUMERICAL TECHNIQUES FOR TOOL AND FIXTURE DESIGN

Out of several different numerical methods, the finite element method (FEM) can be best utilized to analyze different types of horn and booster geometries. It can solve a practical problem with necessary boundary conditions, which are tough to derive using conventional methods by means of differential equations [4]. For the present investigation, commercially available finite element programs ANSYS®, ABAQUS®, and COMSOL® are used to find out the resonant frequencies, the safe working stresses, the temperatures at high-stress regions, the nodal displacement and the knurl indentation of an acoustic horn and booster with the required dimensions.When a structure is actuated by dynamic loading, it is sensible to assume that its response will vary. In such cases, the dynamic analysis may have to be executed to determine both varying loads and responses. Thus, to investigate the effect of cyclic loading on both components, a dynamic analysis is conducted in this study. All the structural dynamics problems can be based on D'Alembert's principle, which gives the time-dependent response of every nodal point in the structure by including the "inertia force" and "damping force" terms in the equation. Therefore, the general equation of motion can be presented by [5]

$$[M].\left\{\frac{\partial^2 \xi}{\partial t^2}\right\}+[C]\left\{\frac{\partial \xi}{\partial t}\right\}+[K].\{\xi\}=\{F\} \tag{3.42}$$

The first step of the dynamic analysis is achieved by performing the *modal analysis*. This analysis is necessary to obtain the natural frequencies and appropriate mode shapes of the horn and booster in such a way that they will be resonating with other systems. Meanwhile, it is assumed that the materials used for both elements have little damping properties; therefore, the equation of motion for a undamped system can be written as

$$[M].\left\{\frac{\partial^2 \xi}{\partial t^2}\right\}+[K].\{\xi\}=\{0\} \tag{3.43}$$

For a free-free vibrated linear system, it is assumed that each node has a sinusoidal function of the peak amplitude of that node, and the displacement vector can be written as

$$\{\xi\}=\{A\}\sin(\omega t) \tag{3.44}$$

The velocity vector can be written as

$$\{\dot{\xi}\}=\{A\}.\omega.\cos(\omega t) \tag{3.45}$$

Similarly, the acceleration vector can be expressed as

$$\{\ddot{\xi}\}=-\{A\}.\omega^2.\sin(\omega t) \tag{3.46}$$

Substituting the values of Eqs. (3.44) and (3.46) in Eq. (3.43), we get

$$[M].[-\{A\}.\omega^2.\sin(\omega t)]+[K].[\{A\}\sin(\omega t)] = \{0\} \tag{3.47}$$

$$\sin(\omega t)[-[M].\{A\}.\omega^2+[K].\{A\}] = \{0\} \tag{3.48}$$

The different modal properties of both elements can be determined from the above equation (Eq. 48). The Eigenvalue equation can be represented as

$$([K].\lambda_E[M])\{A\} = 0 \tag{3.49}$$

It is important to note that the total numbers of Eigenvalues or natural frequencies are equal to the total number of degrees of freedom in the model, and each of them corresponds to an Eigenvector or mode shape. It is worth noting that, usually, the first few Eigenvalues of the model are preferred. Because the FEM gives an approximate solution to the problem, the higher Eigenvalues and vectors are inaccurate.

The second step of the dynamic analysis is the *harmonic analysis*, which is required to find out the prolonged cyclic response of both the elements under periodic loading. This analysis is also known as *frequency response analysis*. It solves D'Alembert's principle shown in Eq. (3.42) for linear structures undergoing steady-state vibrations. It is assumed that all the points/nodes are moving with the same known frequency. But, due to the presence of damping, phase shift may occur. The nodal displacements of in-phase particles are given by

$$\{\xi\} = \{\xi_{max}\}e^{i\omega t} \tag{3.50}$$

The velocity is given by

$$\{\dot{\xi}\} = \{\xi_{max}\}i\omega.e^{i\omega t} \tag{3.51}$$

The acceleration can be written as

$$\{\ddot{\xi}\} = -\{\xi_{max}\}\omega^2.e^{i\omega t} \tag{3.52}$$

Likewise, the force can be defined as

$$\{F\} = \{F_e\}e^{i\omega t} \tag{3.53}$$

Putting the values of Eqs. (3.50) to (3.53) in Eq. (3.42), we get

$$[M].-\{\xi_{max}\}\omega^2.e^{i\omega t}+[C].\{\xi_{max}\}i\omega.e^{i\omega t}+[K].\{\xi_{max}\}e^{i\omega t} = \{F_0\}.e^{i\omega t} \tag{3.54}$$

$$(-\omega^2.[M]+i\omega.[C]+[K])\{\xi_{max}\} = \{F_0\} \tag{3.55}$$

The inertia and static loads on the nodes of each element can be computed. The values for these two loads are represented as follows:
  For inertia loads

$$\{F^m\}_o = (2\pi\omega)^2 [M_e].\{\xi\}_e \qquad (3.56)$$

For static loads

$$\{F^K\}_o = -[K_e].\{\xi\}_e \qquad (3.57)$$

The third step is the *fatigue analysis* of both horn and booster. The objective of this study is to characterize the capability of a material to persist through the cyclic loads it may experience during its lifetime. Because ultrasonic vibration occurs at 20,000 cycles/sec, the component that bears this load over a year can experience high cycle fatigue (HCF). However, the fatigue loading that causes the maximum damage cannot be easily seen. To keep up the trail of loading occurrences for a particular node, the maximum stress intensity range is calculated by a "rain flow" or "range counting" method. These stress differences and intensity are calculated as

$$\{\sigma\}_{i,j} = \{\sigma\}_i - \{\sigma\}_j \qquad (3.58)$$

The stress intensity is calculated based on $\{\sigma\}_{i,j}$ and it is represented as

$$\sigma_1(i,j) = Max\left(|\sigma_1 - \sigma_2|, |\sigma_2 - \sigma_3|, |\sigma_3 - \sigma_1|\right) \qquad (3.59)$$

The interim maximum alternating shear stress is calculated and represented as

$$\sigma_{i,j}^d = \frac{\sigma_1(i,j)}{2} \qquad (3.60)$$

Eventually, the maximum alternating shear stress is calculated as

$$\sigma_{i,j}^c = K_e.\sigma_{i,j}^d \qquad (3.61)$$

The rise in temperature due to the fatigue loading on both components is the concluding analysis. It is based on the principle that when a solid is subjected to elastic cyclic stresses, the material is heated due to alternate expansion and compression. When the alternating pressure is released, then it returns to its initial shape and temperature. By utilizing the Galerkin technique of finite element analysis, the governing equation can be represented as

$$[C].\{T\} + [K].\{T\} = \{Q\} \qquad (3.62)$$

Finite element analysis (FEA) is one such promising numerical method that can be used to solve complex problems. In the ultrasonic field, FEA has been applied to determine the vibrational characteristics of the modeled tools prior to their

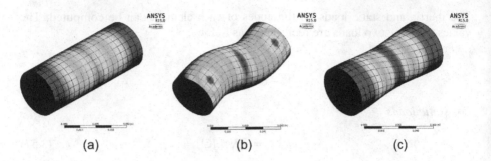

(a)   (b)   (c)

FIGURE 3.2 Classifications of mode shapes with contours: (a) longitudinal mode, (b) bending mode, (c) torsional mode.

application in the manufacturing industries [3]. It uses a mesh of elements to connect the nodes and may also be used for modeling the structure. It requires the material properties along with the loading conditions to do a simulation, and the results can be compared with the analytical results. The FEA software ANSYS® has been used to determine the natural frequencies and corresponding mode shapes of a model [6]. In general, three mode shapes are observed during the analysis of horn and booster: bending, torsion, and longitudinal modes. Figure 3.2 shows these shapes after a finite element (FE) simulation of the rod with a uniform cross-section area as an illustration.

This chapter explains the design, modeling, and fabrication of ultrasonic spot welding components such as the horn, booster, and anvil for a lateral drive system. The horn and booster are the major components that are responsible for transferring the ultrasonic energy to the weld zone [7]. Thus, these components are necessary to build a welding system that will produce high-quality joints, and in this regard FEA is very useful. This analysis aims to get a tuned horn and booster at a particular frequency with low-stress concentration and high magnification. Subsequently, these components are manufactured to the specifications derived from the FEA.

Usually, the ultrasonic generator produces 20 kHz of longitudinal frequency. To ensure the required performance from the welding horn, it is necessary to isolate the longitudinal mode frequency from other modes. Notably, much attention is given to design of the horn, necause the horn acts as a tool, and it wears very fast as compared to the booster. Thus, a horn with multiple weldable tips should be designed to be made from high-wear-resistance and low-acoustic-loss material [8]. Initially, the lengths of both components can be calculated theoretically from Eq. (3.27) using particular material properties, as shown in Table 3.1 [9]. Generally, D2 steel and titanium (Ti-6Al-4V) are used for horn and booster materials respectively for 20 kHz frequency, and their theoretical lengths are found to be 130 mm and 126.8 mm, respectively. These components are designed to be half of the wavelength to match the frequency of the ultrasonic generator. Modifying the dimensions of input and output working surfaces can lead to the change in vibration amplitude. Thus, the performance of the ultrasonic system is measured by its amplification ratio/gain, as described in Eq. (3.41). It represents

TABLE 3.1

Properties of Horn (D2 Steel) and Booster (Ti-6Al-4V) Materials

Property	Units	D2 steel	Ti-6Al-4V
Density (ρ)	Kg/m^3	7670	4430
Young's modulus (E)	GPa	210	114
Poisson's ratio (ν)		0.3	0.33
Velocity of sound (C)	m/sec	5232	5072

the capability of the system to weld high thickness materials. The dimensions of the input and output ends are fixed by the size of the booster and the particular weld tip size. For the illustration purpose, the input and output diameters of the horn are considered as 44 mm and 35 mm, and for the booster, these are 55 mm and 44 mm, respectively. The spot-welding horn with the booster is presented in Figure 3.3.

The supports that hold the horn and booster are called support rings/flanges. These are attached to the nodal sections and should have larger diameters than the horn and booster. Otherwise, there will be a damping of amplitude in the latter stage. Another advantage of designing these integral components is that one can figure out how to apply the clamping force perpendicular to the weld specimens.

The welding tip is intended to be an integral part of the USW system to eliminate the stress concentration, increase the ultrasonic energy intensity at the welding zone, and reduce the slippage between the contact surfaces of horn tip and upper specimen. The weld tip surface has knurled patterns to engage into the specimens effectively.

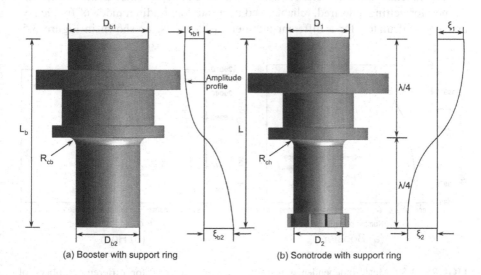

(a) Booster with support ring   (b) Sonotrode with support ring

FIGURE 3.3   Modeled booster and horn with amplitude profiling.

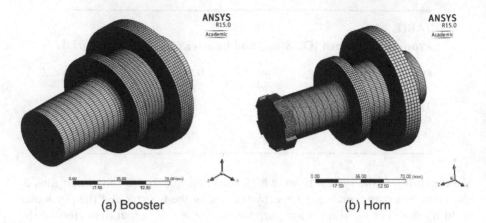

(a) Booster   (b) Horn

FIGURE 3.4   Meshed 3D models for finite element analysis.

The dimensions of the knurl patterns are comparably less than the area of the weld tip. For carrying out FE simulation, one must choose element type, real constant, material properties, geometry, meshing, boundary conditions, etc. judiciously. As it is a three-dimensional analysis, the solid187 element has been selected, which can analyze irregular shapes of both components, and also has better compatibility to metal deformation. It can also be applied to the curved boundary with plasticity, creeping, expansion, large deformation, and even a body with failure. In FEM, a variety of mesh techniques exists to mesh models of different topology. The horn and booster which are being investigated are discretized by a mesh of uniform and straightforward hexahedral elements connected by several nodes, as seen in Figure 3.4. These elements have been chosen because of they are more accurate in measuring higher-stress gradients than triangular elements. The mesh independence test has been performed for getting a desired, reliable, and accurate longitudinal mode of frequency, and it is calculated for a different number of elements, as shown in Figure 3.5.

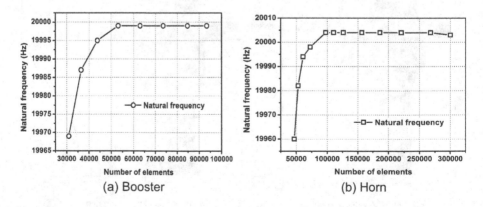

(a) Booster   (b) Horn

FIGURE 3.5   Mesh independence test of natural frequency for different numbers of elements.

A sensible number of elements (i.e., 97,368 for horn and 53,135 for the booster) are selected, keeping in view the necessary computational time, the convergence of the tuned mode, and the accuracy of the result. It shows that the present model is sufficiently accurate to solve the problems of frequency domain for these two components. Thus, it provides confidence to the welder for the determination of natural frequencies and mode shapes.

Modal analysis through the block Lanczos method is used to determine the natural frequencies and mode shapes of the acoustic horn and booster. It is a linear analysis, and any nonlinearities such as plasticity and contact elements are ignored, as those effects are insignificant. It ensures the rigidity of the whole structure. The lower the natural frequency, the lesser the rigidity of these components. With proper design, the horn will resonate with other parts of the system in the range of 19.5-20.5 kHz using Eq. (3.49). There is a possibility of getting more than one natural frequency and mode shape within this range. Thus, different mode shapes are observed, such as longitudinal, bending, and torsional. However, the first and longitudinal or axial mode shape is always favored, for two reasons: firstly, the transducer produces mechanical vibration in the longitudinal direction; and secondly, when modes 2 and 3 are superimposed with the primary longitudinal mode of ultrasonic vibration, it causes a bending mode. This is an undesirable phenomenon which can jeopardize the advantages of ultrasonic vibration, causing poor welding operation. Figure 3.6 depicts the natural longitudinal mode of vibration.

Several trial runs have been done for shortening of length and adding radius of curvature at the nodal plane of the components so that the frequency of the axial mode should match the machine frequency, and the stress concentration would be as low as possible. Firstly, from the modal analysis, the natural frequencies of horn and booster are uncovered by adjusting the length of both parts. Subsequently, they are shortened gradually by 0.5 mm at a time, keeping all other dimensions constant to get the desired 20 kHz frequency. During these analyses, bending, torsional, and longitudinal mode shapes are obtained. As stated earlier, these mode shapes are unwanted phenomena, and they gradually disappear when the total lengths of the booster and horn are set to 125.4 mm and 126.8 mm, respectively. The various lengths of both components with the first longitudinal vibration mode are shown in Figure 3.7. It is observed that the natural frequencies of both the parts modeled by FEA approach the operating frequency of the system, after shortening of the length.

The radius of curvatures (R_{cb} and R_{ch}) also has been changed to account for the danger of stress concentration, which is applied to the junction of the big and small cross-section for both the components. Figure 3.8 shows the effect of changing R on the natural frequency. It has been observed that with increasing R, the natural frequency also increases. When the R values are 4.5 mm and 5.5 mm, then the 20,004 Hz and 19,999 Hz frequencies are obtained for horn and booster, respectively. Hence, it is observed that the natural frequencies of the horns modeled by the FEA approach the operating frequency of the system.

Frequency/harmonic response analysis is a technique used to determine the steady-state response of a linear structure to loads that may vary sinusoidally with time. It involves a time history processor, which shows the stresses at different

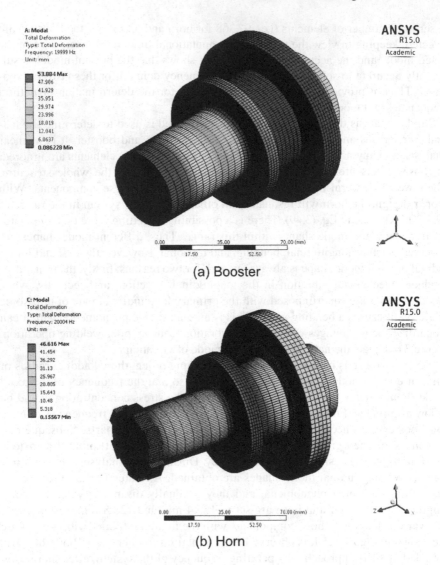

A: Modal
Total Deformation
Type: Total Deformation
Frequency: 19999 Hz
Unit: mm

53.884 Max
47.906
41.929
35.951
29.974
23.996
18.019
12.041
6.0637
0.086228 Min

ANSYS
R15.0
Academic

0.00   35.00   70.00 (mm)
     17.50   52.50

(a) Booster

C: Modal
Total Deformation
Type: Total Deformation
Frequency: 20004 Hz
Unit: mm

46.616 Max
41.454
36.292
31.13
25.967
20.805
15.643
10.48
5.318
0.15567 Min

ANSYS
R15.0
Academic

0.00   35.00   70.00 (mm)
     17.50   52.50

(b) Horn

FIGURE 3.6   Natural frequencies from modal analysis.

positions of the parts, and displacement of nodes at the booster output end as well as at the free end of the horn. Such an analysis has been carried out without considering the damping effect, due to the modest damping capacity of horn and booster materials. Therefore, the damping term in Eq. (3.55) is neglected. The ultrasonic generator can produce a maximum amplitude of 30 μm per its specifications. By using Eq. (3.41), the amplification factor for both the components is determined as 1.5. Thus, the theoretical output amplitude of the booster and horn is modified to 45 μm and 67.5 μm for the first longitudinal vibration frequency. The results obtained from the modal analysis are taken as the input for the harmonic analysis. The graphical

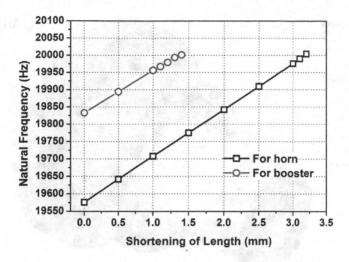

FIGURE 3.7 Variation of natural frequencies with shortening of lengths from theoretical ones [10].

representation of amplitude distribution is displayed in Figure 3.9. The working amplitudes of the booster output end and tool tip are found to be 47.35 μm and 68.86 μm, respectively, when the forced frequency matches with the first longitudinal natural frequency of both the parts.

The nodal planes are located at the midsection of the horn and booster, which represent a good agreement with the theoretical principles (Figure 3.10). The present analysis also reveals stresses at different positions of the horn and booster for the frequency given by modal analysis. The stress plots for both components are shown

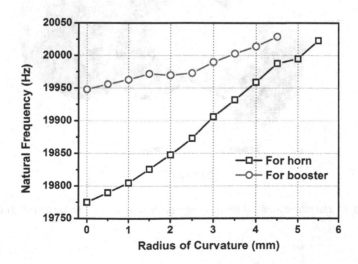

FIGURE 3.8 Variation of natural frequencies with radius of curvatures [10].

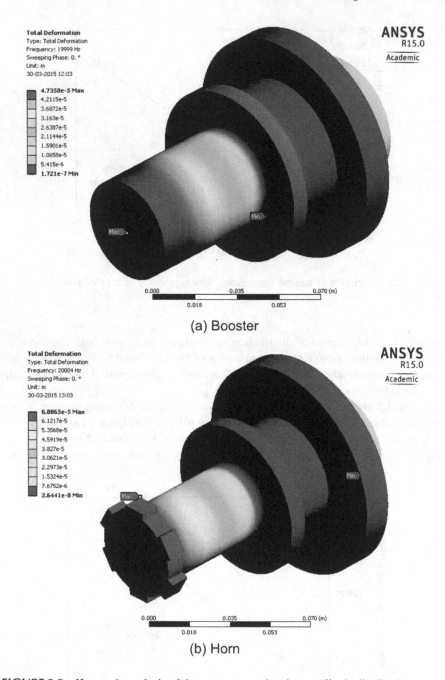

(a) Booster

(b) Horn

FIGURE 3.9   Harmonic analysis of the components showing amplitude distribution.

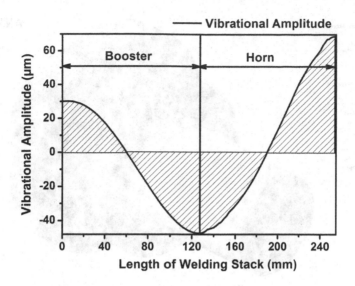

FIGURE 3.10   Amplitude variation plot with respect to length of both parts.

in Figure 3.11. These indicate that both parts experienced a sinusoidal waveform at the same frequency as the input. According to the stress profile, it increases with time up to the nodal position and then gradually fades away. Therefore, the values of stress are found to be minimal at both the free ends, thus also satisfying the theoretical principles.

After application of the loading condition connected with the longitudinal mode of vibration, the stress of the horn can be analyzed using von Mises criteria. Figure 3.12 illustrates the von Mises stress distribution along the length of the welding stack. The contour plots show that it increases with the gradual increase in the length, and becomes maximum at the abruptly changing sections. The maximum stresses observed in this study are 196.67 MPa and 576.6 MPa for booster and horn, respectively. In this study, the maximum stresses of the horn and booster design are significantly lower than the yield strength of the D2 steel. Hence, the ultrasonic welding booster and horn can meet safe operating conditions in terms of stress levels.

Amplitude uniformity is defined as the variation between the minimum and maximum amplitude at the weld tip surface of the horn. In USMW, the quality of the joint depends on the uniformity of amplitude. Figure 3.13 reveals the normalized amplitude (ratio of vibrational amplitudes at different points on the weld tip to the maximum amplitude observed on the weld tip) of the weld tip surface. From this figure, about 99.8% of uniformity in amplitude is obtained.

According to ASTM [11], *fatigue* is defined as the phenomenon associated with the gradual and localized development of permanent changes within the structural body due to the application of repeated stress or strain. In the present context, the horn and booster are subjected to cyclic loading of 20,000 cycles/sec. It is projected that the components will be working 10 hours per day at a rate that produces six welds per minute. Hence, in a year, more than 10^5 welds are produced, if the components

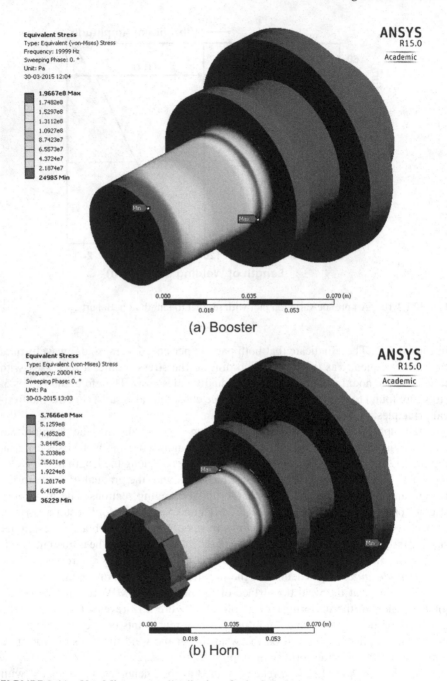

FIGURE 3.11 Von Mises stress distributions for both components.

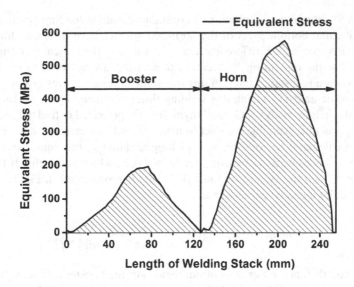

FIGURE 3.12   Stress variation plot with respect to length of both parts [10].

remain in working condition throughout the year as expected, which creates a situation of high cycle fatigue. HCF affects the entire life of the tools.

Furthermore, it is tough to pinpoint the initiation and proliferation of cracks. Projections are based on the S-N (stress-cycle) curve. The relevant horn and booster material data are obtained from the military handbook [12]. In a conventional USMW operation, the maximum and minimum stress levels are changed or varied according to the loading and unloading conditions. Thus, these types of problems are solved by

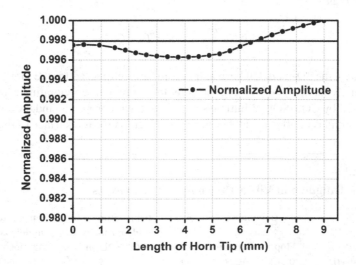

FIGURE 3.13   Variation of FE simulated normalized amplitude of horn tip with horn tip length excited at 20.04 kHz.

the ANSYS® fatigue module with non-constant amplitude loading conditions. The top, middle, and bottom parts of the horn and booster have different dimensions, so these parts experience different stress values during the typical welding time of 1 second. For the remaining 59 seconds, these parts are not undergoing any load (that is, they are in a static state). In this study, the equivalent stress values obtained from harmonic analysis during this welding time are taken as the maximum input values to the fatigue analysis. Gerber's hypothesis is preferred to find out mean stress effects, as it is quite suitable for ductile material and can consider its −ve and +ve values. To calculate the total amount of fatigue damage, the cumulative damage calculations should be done by using the Miner rule, which states that if this value is equal to 1, then the system will fail [13]. This is represented in Figure 3.14, and mathematically it can be expressed as

$$Cumulative\ damage(C_d) = \frac{Design\ life(N_i)}{Available\ life(N_f)} \tag{3.63}$$

It is observed that, at the top and bottom regions of the booster and horn, the cumulative damages are 0.142 and 0.218, respectively, which are well below the limit. However, cumulative damages of 1.175 and 1.111 are found in the nodal regions, where the stresses are high, suggesting the possibility of failure of both parts at given working frequencies and amplitudes. The results are summarized in Table 3.2.

During fatigue loading, the components are under repeated elastic cyclic stresses, and the ultrasonic wave propagates through them. For this reason, heat generation will take place due to the elastic deformation of the material. But, when the pressure is released, the material comes back to its actual shape and initial temperature per the principle of vibro-thermographic analysis. To simulate the temperature rise in the specimen, the Dulong-Petit law of the heat transfer equation is used, and is represented as

$$\rho.C_p.\frac{\partial T}{\partial t} - \nabla(K.\nabla T) = Q_h \tag{3.64}$$

In this study, the 3D-modeled horn and booster are considered for the vibro-thermographic analysis; the gradual elevation of temperature in those vibrating components is simulated by COMSOL® Multiphysics software with proper boundary conditions. Input amplitudes of 30 μm and 45 μm are provided on the top surface of the booster

---

TABLE 3.2
Results of Fatigue and Vibro-Thermographic Analyses

Components	Cumulative damage			Temperature at the middle zone from COMSOL® in °C	Temperature at the middle zone from thermocouple in °C
	Top	Bottom	Middle		
Horn (D2 steel)	0.218	0.218	1.111	63.5	64.85
Booster (Ti-6Al-4V)	0.142	0.142	1.175	49.8	50.24

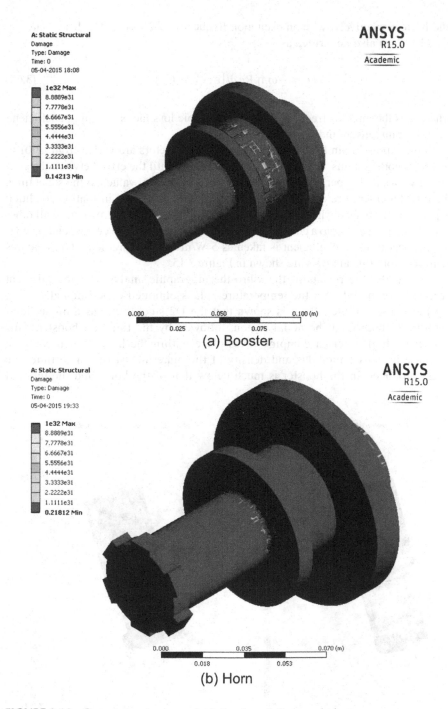

FIGURE 3.14 Cumulative damage calculation from fatigue analysis.

and horn, respectively, with an excitation frequency of 20 kHz. The heat source in Eq. (3.64) can also be written as

$$Q_h = \frac{1}{2}\omega.\eta.\operatorname{Re}al\left[\varepsilon:Conj(C:\varepsilon)\right] \tag{3.65}$$

where ω is the angular frequency, η is the isotropic loss factor/damping coefficient, ε is the strain tensor, and C is the elasticity tensor.

In the expression in Eq. (3.65), the isotropic loss factors are varied from 0.001 to 0.008 for both systems. These factors are calculated until the error between the simulated values and experimental results reaches a minimum value. At the same time, the initial temperature is assumed as 293.15 K with a thermal insulation condition applied to the surface where the load is applied. Other than these two ends, all other boundaries are subject to a convective cooling condition, and the corresponding convective heat transfer coefficient is taken as 5 W/m² K. The predicted temperatures achieved from this analysis are shown in Figure 3.15.

Meanwhile, the results of the vibro-thermographic analyses of the different parts are compared with the temperature values obtained experimentally using thermocouples (Table 3.2). It is shown that the D2 steel horn has a higher temperature of 63.5°C at the nodal region, followed by the titanium booster. This is because high internal damping takes place within the horn, which causes a decrease in elastic modulus and density of the material. At the same time, the stress produced in the booster is much below that of the horn, and the internal

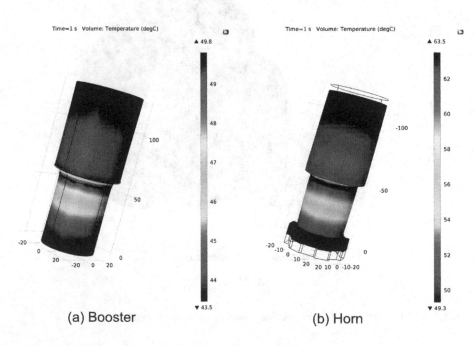

FIGURE 3.15   Predicted temperatures from vibro-thermographic analysis [10].

damping coefficient is lesser as compared to that of the horn. For this reason, the vibrations can easily be transmitted through it to the horn for the effective welding of thin metal sheets. Meantime, errors of 2.08% and 0.87% are obtained for horn and booster when the experimental temperature values are compared with the simulated results.

Designing the welding stack (composed of the horn and the booster) is a challenging task because the high-frequency vibration should precisely occur at the welding tip, and the amplitude should be maximized at the welding tip. In this analysis, both of these components have been investigated simultaneously instead of being considered separately. From the harmonic analysis, it may be observed that the input amplitude is modified by 1.509 times for the booster, and again it is amplified by 1.5 times for the horn to create a good weld between the sheets. Thus, at the end of the horn, 67.92 µm amplitude is obtained, and this value is quite close to the theoretical result. Also, at the nodal position of the booster and horn—i.e., at 65.4 mm and 192.2 mm—the amplitude becomes zero, and the von Mises stresses turn out to be 187.7 MPa and 563.08 MPa respectively. These variations in amplitude and stresses with the length of welding stack are given in Figure 3.16. The summary of the results (given in Table 3.3) shows that there are 1.11% and 2.52% errors in the length calculation of both systems. Similarly, a 0.61% error is obtained for both systems while calculating the magnification ratio.

The FEA should be performed considering the specific geometry of the horn tip to study the plastic deformation and stress distribution at the weld interface. The amount of plastic deformation and material flow under the sonotrode knurl tip are the deciding factors in providing sound weld joints in the bonding process. However, it cannot be predicted by theoretical analysis. The design of the gripping of the horn tip has a significant impact on the plastic deformation and heat generation between the weld specimens, and it ultimately affects the overall performance of the ultrasonically welded joint. In this regard, the Johnson-Cook plasticity model is the most suitable method for modeling the high strain rate deformation of the materials. In the USW process, high strain material deformation occurs over a wide range of strain

TABLE 3.3
Comparison Between Theoretical and FEM Results of Booster with Horn

Sl No.	Factors	Theoretical calculated horn	Theoretical calculated booster	FEM analyzed horn	FEM analyzed booster	% error in horn	% error in booster
1	Frequency (kHz)	20	20	20	20	0	0
2	Total length (mm)	130	126.8	126.8	125.4	2.52	1.11
3	Diameter ratio	1.25:1	1.25:1	1.25:1	1.25:1	0	0
4	Magnification factor	1.5	1.5	1.5	1.509	0.61	0.61

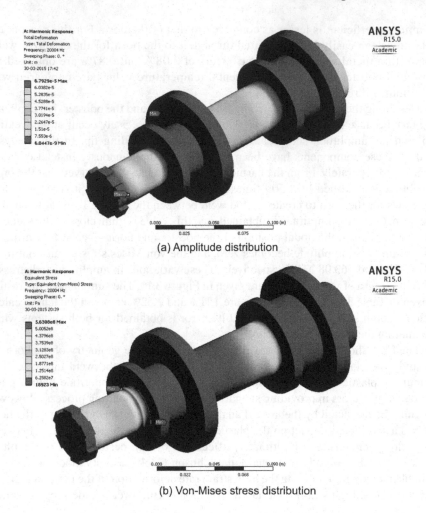

FIGURE 3.16 Simulated results of amplitude and stress distributions for booster with horn.

and temperature. This model includes an isotropic hardening rule in which the effective stress σ can be defined as follows:

$$\sigma = \left[A + B\left(\overline{\varepsilon}^{pl}\right)^n \right]\left[1 + C\ln\left(1 + \dot{\varepsilon}_0\right)\right]\left(1 - \hat{\theta}^m\right) \tag{3.66}$$

where $\dot{\varepsilon}_0$ represents the normalized effective plastic strain rate (typically 1.0 s^{-1}); A, B, C, n, and m are the material-related parameters measured at or below the transition temperature; and $\hat{\theta}$ is a nondimensional temperature element which can be defined as

$$\hat{\theta} \equiv \begin{cases} 0 & \textit{for } \theta < \theta_{transition} \\ \left(\theta - \theta_{transition}\right)/\left(\theta_{melt} - \theta_{transition}\right) & \textit{for } \theta_{transition} \leq \theta \leq \theta_{melt} \\ 1 & \textit{for } \theta > \theta_{melt} \end{cases} \tag{3.67}$$

TABLE 3.4

Material Properties of Cu and Al for Johnson-Cook Model

Material	θ_{melt} (°C)	A (MPa)	B (MPa)	C	n	m
Al	620	148.4	345.5	0.001	0.183	0.859
Cu	1083	90	292	0.025	0.31	1.09

In this expression, θ is the temperature at a particular instant, θ_{melt} is the melting point of the material, and $\theta_{transition}$ is the transition temperature below which there is no dependency of temperature on the yield stress. Thus, all the material-related properties should be considered at or above the transition temperature. As an illustration, this analysis is considered for the USW of Cu and Al sheets. The parameters of the Johnson-Cook model are drawn from Gupta et al. [14] and Johnson and Cook [15], and are presented in Table 3.4.

ABAQUS is employed to analyze this coupled thermo-mechanical FE model for the USW of Cu and Al. The FEA involves two steps. The first step describes the application of the clamping force by the sonotrode tip on the upper specimen for a short duration of time. In this step, the heat generation is ignored. During this time, no dynamic effect is considered, and it is assumed that the interfaces of the weld specimen, sonotrode tip, and anvil surface are intact with each other. This clamping period is relatively short, in comparison to weld time, to reduce the operational and calculation costs. The second step involves the welding time when the joining of the specimen has happened. Figure 3.17(a) shows a cross-sectioned FE model, which is comprised of weld specimens (i.e., Al and Cu), sonotrode, and anvil. It is assumed that the width and length of the specimens are equal to the anvil dimensions, and only a portion of the sonotrode is considered for this analysis. Figure 3.17(b) is a magnified view of the weld cross-section, where the various types of meshing can be observed. The denser mesh grids are at the contact interfaces of all elements to reduce the hourglassing effect and improve the computational efficiency. However, the parts away from the weld cross-section have a sparse type of meshing, again to reduce the calculation time. The sonotrode and anvil are treated as rigid bodies due to their higher stiffness and relatively less deformation than the weld specimen. Nevertheless, these components can transmit thermal response during the USW process. The eight-node trilinear displacement and temperature element (C3D8RT) with reduced integration point and hourglass control technique is considered for the coupled thermo-mechanical analysis of the process. In this element, C stands for continuum, 3 for trilinear, D for displacement, R for reduced, and T for temperature. The reduced integration technique can save a lot of calculation time, and the hourglass control ensures the convergence of the model by preventing heavy distortion of the elements.

According to the design of the sonotrode knurls, the stress distribution is uneven at the different parts of the weld interface. This phenomenon significantly influences the bonding mechanism in the USW process. During the USW process, as the weld materials move relative to each other, shear stress is developed between them due to the friction. Figure 3.18(a) is a schematic diagram of the clamping

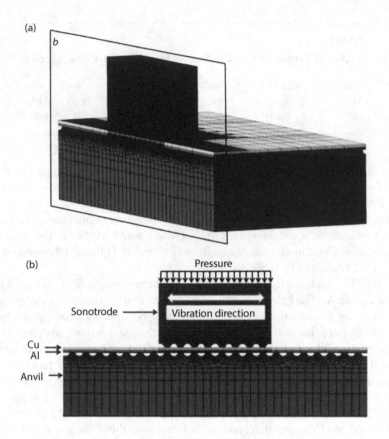

FIGURE 3.17 FE model of weld specimen in the USW process: (a) half cross-section model, (b) magnified view of weld interface [16].

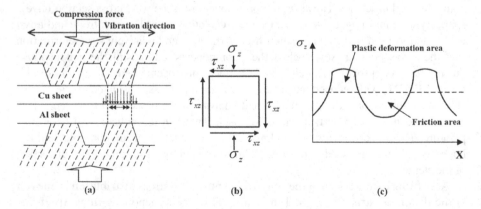

FIGURE 3.18 Schematic diagram of plastic deformation and interface friction between the weld specimen: (a) clamping force and shear stress direction, (b) direct and shear stress distribution under sonotrode tip, (c) areas of plastic deformation and friction [17].

force and stress at the interface of the Cu and Al materials. Figure 3.18(b) exhibits the direction of the external force and the shear stresses. This shear stress ($\tau_{xz} = \mu_f \times \sigma_z$) can be calculated assuming the dry friction condition. μ_f represents the coefficient of friction and σ_z is the direct stress applied due to the external force. It is well understood that the amount of plastic deformation depends on the direct stress, and this amount can be obtained from the von Mises yield criteria. This can be represented as

$$\sigma_z \le \left(\sigma_s / \sqrt{1 + 3\mu_f^2} \right) \tag{3.68}$$

σ_s is the yield strength of the weld specimen. Initially, the part under the sonotrode tip endures higher compressive force, and plastic deformation occurs in this zone. Meanwhile, the interface friction mainly occurs in these areas as the welding starts, as shown in Figure 3.18(c). During the welding process, the yield strength gradually decreases with respect to time due to the combined effect of thermal and acoustic softening. The plastically deformed area is gradually increased and the frictional area is decreased, resulting in high strength and a reliable joint.

Figure 3.19 demonstrates the comparative numerical models of two types of sonotrode knurl patterns: flat tip and serrated tip. There are eight parallel serrations on the sonotrode tip, and the pressure distribution is asymmetrical in the welding zone, as illustrated in Figure 3.19(a). This contact pressure is higher under the sonotrode knurls than between the knurls. In this type of sonotrode knurl pattern, the pressure value is reduced as the distance from the center of the sonotrode ridges increases. Meanwhile, the model with a flat sonotrode tip exhibits uniform pressure distribution at the center of the weld zone with moderate variations at the periphery of the weld spot (Figure 3.19(b)). However, the contact pressure produced by the serrated sonotrode tip is much higher than that of the flat sonotrode tip during the USW process.

The details of the structural deformation under the sonotrode knurl pattern at the end of the USW process is presented in Figure 3.20(a). The contact region of weld specimens, along with sonotrode and anvil surfaces, is magnified for further analysis (Figure 3.20(b)). At the end of the USW process, the interface temperature softens the weld samples resulting in the more effortless penetration of the sonotrode ridges to the top weld specimen and occurrence of the material flow at the weld interface. Meanwhile, it is also observed that the deformation is not symmetrical about the weld center. The lower specimen deformed less than the top specimen. Thus, there is a need for good anvil design, which will provide the necessary support for plastic deformation during the USW process.

Figure 3.21 reveals the effective von Mises stress distribution on the top and bottom weld specimens during the USW process. Interestingly, one can see that the stress on the top specimen is lower than that on the bottom specimen. This is because the highly intensified ultrasonic energy is concentrated on the upper specimen, resulting in more acoustic softening of that sheet. However, this acoustic softening effect is less on the bottom sheet.

The temperature distribution during the USW process is simulated in two states: during welding time and during the cooling time (Figure 3.22). The temperature

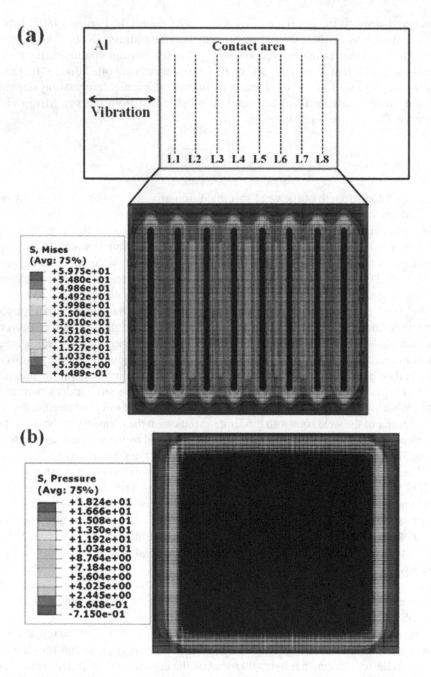

FIGURE 3.19 Pressure distribution profiles on Al weld surface during clamping process under (a) serrated knurl edges and (b) a flat sonotrode edge [18].

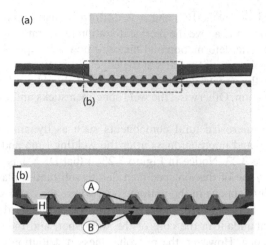

FIGURE 3.20 (a) Overview of the deformed shape of the specimen at the end of the USW process; (b) magnified view of dot-outlined portion, where A and B represent the Cu and Al specimens respectively, with H indicating height [18].

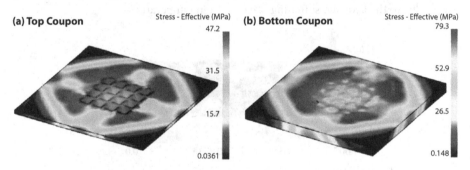

FIGURE 3.21 Effective von Mises stress distribution on (a) top specimen and (b) bottom specimen [19].

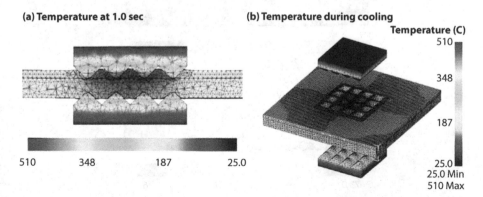

FIGURE 3.22 Temperature distribution during the USW process (a) at the end of welding time and (b) during the cooling process [19].

in the weld zone is about 500°C, and it is uniformly distributed over the surface area. This temperature is above the recrystallization temperature of weld materials. Thus, sufficient plastic deformation and material flow are expected. Figure 3.22(b) reveals that the cooling initially starts at the edge and moves to the core of the weld zone. Meanwhile, the sonotrode and anvil materials should not be affected by this temperature generation. Otherwise, the weld specimen sticks and wear of these components occurs.

The simulated microstructural components such as dynamic recrystallization (DRX), grain size, and microhardness after the welding time and cooling step are presented in Figure 3.23. Notice in Figure 3.23(a) that DRX has occurred in most parts of the weld zone of the top specimen due to substantial plastic deformation, and that DRX only happens in the bottom sheet where the anvil knurls indent this coupon. Likewise, Figure 3.23(b) shows the grain size of the weld interface. Larger grains are more abundant in the core of the weld spot, and then become finer in the surrounding zone. However, the microhardness distribution exhibits the exact reverse trend of the grain size (Figure 3.23(c)). The core of the weld spot has the coarser grain size, and the surrounding zone has the finer grain size. Thus, the core has a lower microhardness than the surrounding area. This low hardness of the core is also due to the acoustic softening effect at high temperatures.

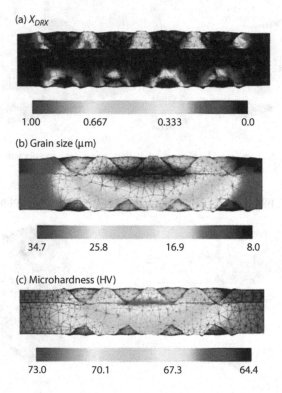

(a) X_{DRX}

1.00   0.667   0.333   0.0

(b) Grain size (μm)

34.7   25.8   16.9   8.0

(c) Microhardness (HV)

73.0   70.1   67.3   64.4

FIGURE 3.23 Dynamic recrystallization, grain size, and microhardness distribution during the USW process [19].

3.3 INFLUENCE OF TOOL AND FIXTURE DESIGN ON WELD STRENGTH

Most studies concentrate on the effects of the different process parameters on the weld strength and temperature generation during the USW process. Because ultrasonic welding is a complex process, it is necessary to determine the quality of the weld in terms of fracture pattern, wake features, and flow behavior at the plastic deformation zone. The design of ultrasonic tooling plays a significant role in the determination of quality and strength of the welding. A good design of this part provides proper friction between the weld coupons and avoids slippage between the sonotrode and top specimen. Therefore, it is necessary to examine the weld strengths, effects of welding parameters, and weld quality characteristics using different anvil geometries.

For a welder who wants to achieve excellent quality in welding, it is essential to use suitable process parameters and proper design of anvil surface patterns. The specifications of anvil knurl patterns are given in Table 3.5.

The tensile shear and T-peel failure load curves as a function of weld time for these three anvil patterns are shown in Figure 3.24. These two strengths increase with weld time up to 0.75 sec. After this, the strengths gradually decrease due to

TABLE 3.5
Details of Anvil Tool Geometries with Specifications (Scale in Mm)

Anvil cap no.	1	2	3
Material	SS 304	SS 304	D2 steel
Cutting width (mm)	0.55	0.55	0.55
Noncutting width (mm)	1	1	0.65
Angle(degrees)	75	45	75
Shape	Parallelogram	Parallelogram	Truncated pyramid
Textures			

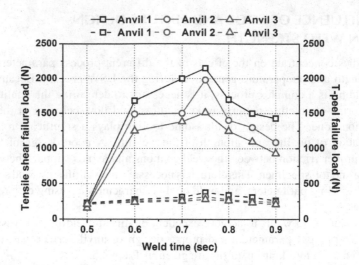

FIGURE 3.24 Tensile shear and T-peel failure loads of Al-Cu weld coupons with weld time for various anvil patterns.

severe plastic deformation and cracking around the weld zone. In the tensile shear tests, the welds generally fractured at the weld interface when the weld time was less than 0.6 sec. The problem may be due to the improper mixing of metal at the plastic deformation zone. However, when the weld time is 0.7 or 0.75 sec, the fracture occurs at the periphery of the weld zone, and partial tearing of the aluminum sheets is also observed. Thus, it is considered a good weld condition, and in this zone, the sufficient amount of heat generation and plastic deformation occurs. After 0.75 sec, nugget pullout failure occurs due to the formation of cracks in the weld zone.

Furthermore, there is a trend of increasing tensile shear and T-peel strength with increases in the non-cutting width and angle of the knurls. The surfaces of copper have not been deformed plastically when the smaller non-cutting-width type of knurl is used. For instance, in between anvil 1 and anvil 3, the angle of the knurl is same, but the non-cutting width is different. Another reason for lowering both strengths is the geometrical shape of the knurls.

For the present study, flat-type knurls are engraved on anvils 1 and 2, whereas on anvil 3, pyramidal-type knurls are created. It has been experimentally found for anvil 3 that an insufficient plastic deformation occurred because its shape prevents the further indentation of the horn tip into the top surface of the aluminum sheet. In contrast, the larger area of contact and the appropriate angle of anvil 1 make the material flow freely and allow the horn tips to indent further. Also, the part seldom sticks to the sonotrode while using anvil 1. Thus, anvil 1 produced a more significant failure load than anvil 2 or anvil 3. The increased strength values at higher amplitudes are attributed to the immense frictional heat that gives rise to more plastic deformation at the interface layer. At the highest weld time, though, relatively lesser

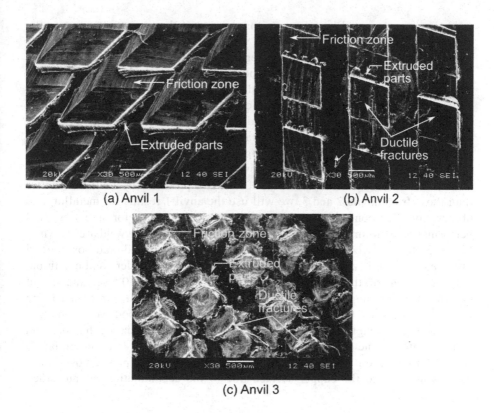

(a) Anvil 1   (b) Anvil 2

(c) Anvil 3

FIGURE 3.25   SEM photographs of impressions on copper sheet by various anvils: (a) anvil 1, (b) anvil 2, (c) anvil 3.

failure loads are obtained due to gradual softening and thinning of the material. Therefore, it can be considered as an overweld condition.

The angle of the knurls also plays a significant role in the generation of higher weld areas with less friction, as evidenced by the SEM photographs in Figure 3.25. In Figure 3.25(a), plastic deformations appear around the knurl patterns of anvil 1; in Figure 3.25(b), plastic deformations can also be observed around the knurl pattern, but these patterns show ductile fractures for anvil 2. Likewise, Figure 3.25(c) depicts severe plastic deformation with fracture patterns near the weld spot. It can be observed from this figure that due to the anvil 3 knurl pattern, the surface of the copper specimen suffered the maximum deformation, and a lot of fractures also occurred. Thus, it is believed that the lowest failure loads are obtained in this condition.

The joint strength of the weldment can be determined by various physical attributes. These weld characteristics are bond density, post-weld thickness, and thermo-mechanically affected zone (TMAZ). This analysis only discusses the various dimensional effects of the anvil on weld strength, and therefore bond density and bond thickness are taken into consideration. Bond density for different weld times is estimated by taking the proportion of the projected region to the entire welded interface region. A decent weld should have thick interfacial bonds without any voids, and these can be observed through microstructural analysis of the weld interface. There are numerous ideas available to describe the formation of a weld in USMW. These ideas include concepts like metallurgical adhesion, which originates from plastic deformation; local heating; and mechanical interlocking. As mechanical interlocking and metallurgical adhesion play a predominant role, cross-sectional views are examined thoroughly here. Three distinct phases are identified during the welding process along the weld cross-section: (1) development of micro bonds, (2) convoluted wake-like features, and (3) flow of Al material to the Cu side. As the joint strengths produced by anvil 1 are better than those from anvils 2 and 3, we will use the anvil 1 results for metallurgical observation. The bond thickness is a vital physical attribute for assessing weld performance. SEM images of under weld, good weld, and over weld are shown in Figure 3.26 for anvil 1. To enhance understanding, the black dotted box marked off in each figure is zoomed in the second column. In an under weld condition, voids are observed; this weld was done at a low weld time of 0.5 sec and also at 68 μm vibration amplitude. To the contrary, a dense bonding region can be noticed in the good weld and over weld conditions. The good weld condition shows a higher strength than the over weld condition, thus indicating that the bond density feature alone cannot adequately describe the effect of dense micro bonds on weld strength. But, in Figure 3.26(b), one can observe the plastic flow of aluminum material to the copper surface due to the accumulation of intensified

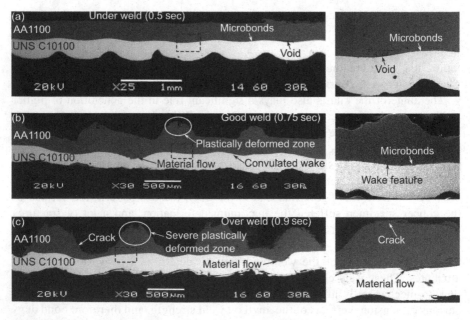

FIGURE 3.26   SEM images of different weld conditions: (a) under weld, (b) good weld, (c) over weld.

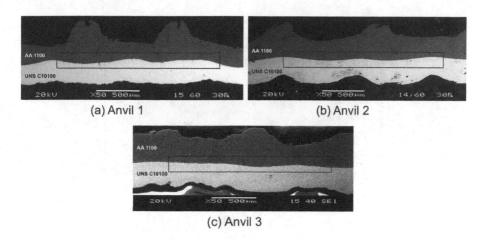

(a) Anvil 1   (b) Anvil 2

(c) Anvil 3

FIGURE 3.27 SEM images revealing wake features for various anvils: (a) anvil 1, (b) anvil 2, (c) anvil 3.

ultrasonic energy. This also depends on the extensive deformation produced by the anvil knurl pattern.

For a weld time of 0.75 sec, the three anvils show a continuous weld interface line, as shown in Figure 3.27. As this figure shows a good weld condition for each anvil, the voids are not seen on the weld line, and penetration of both weld and anvil tip can be observed. Surface extrusion and plastic deformation at the weld periphery of anvil 1 are found to be the largest of all the anvils. Thus, it is one of the reasons for getting higher strength by using anvil 1. Two types of wake features are observed in the morphological analysis of the weld joints. As sufficient plastic deformation occurs at anvil 1, finer wakes are found in the means of the convoluted profile than with the other anvils. A sinusoidal pattern of the interface has also been noticed, which bears a resemblance to anvil tip knurls. In this experiment, the testers tried to find a correlation between interface morphology and geometry of the knurl pattern. It was found that the troughs and crests of the wavy interface are transient and change with the weld time and amplitude. Thus, a consistent relationship between the wake features and the anvil knurl does not exist.

Likewise, Figure 3.28 demonstrates a comparison of joint strengths for serrated and trapezoidal sonotrode edges at various welding times. Here the bonded zone expands in the direction of welding with the increase in weld time. For the serrated-type sonotrode knurls, the plastic flow starts from the top specimen to the weld interface, and it results in growth of the bonded zone, which enhances the joint strength. However, the trapezoidal sonotrode edge shows less plastic deformation and a partially developed bonding zone. Thus, the joint strength is higher with a serrated edge than with a trapezoidal-type edge over the same welding time and weld pressure. Moreover, the normal stress imparted by the serrated sonotrode edge on the weld interface is higher than that generated by the trapezoidal edge, due to the serrated edge's low penetration capability and high contact area which yield the improvement in joint strength.

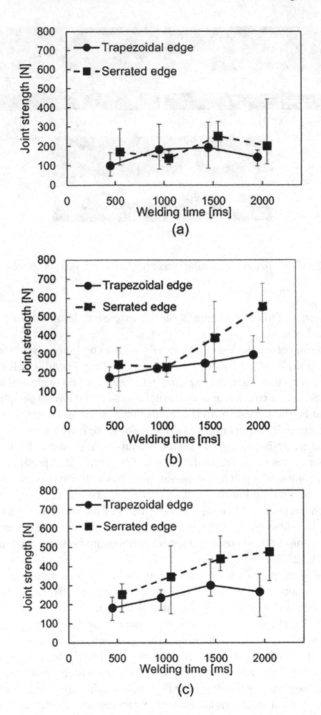

FIGURE 3.28 Comparison of joint strengths produced by trapezoidal and serrated sonotrode edges under normal forces of (a) 294N, (b) 588N, (c) 882N [20].

REFERENCES

1. McCulloch E. Experimental and finite element modelling of ultrasonic cutting of food. University of Glasgow, 2008.
2. Merkulov LG, Yakovlev LA. Propagation and reflection of ultrasonic beams in crystals. Sov Phys Acoust 1962;8:72–7.
3. Cardoni A. Characterising the dynamic response of ultrasonic cutting devices. 2003.
4. Amin SG, Ahmed MHM, Youssef HA. Computer-aided design of acoustic horns for ultrasonic machining using finite-element analysis. J Mater Process Technol 1995;55:254–60. doi: 10.1016/0924-0136(95)02015-2.
5. ANSYS R. 11.0 Documentation, SAS IP 2007.
6. Mechanical A. ANSYS Mechanical APDL Theory Reference; Version 13.0. ANSYS Inc, Canonsbg 2013.
7. Troughton MJ. Handbook of plastics joining: A practical guide. William Andrew; 2008.
8. Phillips AL, Kearns WH, Weisman C. Welding handbook. vol. 2. American Welding Society; 1976.
9. Matweb. Material property data. n.d. www.matweb.com (accessed April 6, 2013).
10. Satpathy MP, Sahoo SK. Experimental and numerical studies on ultrasonic welding of dissimilar metals. Int J Adv Manuf Technol 2017. doi:10.1007/s00170-017-0694-2.
11. Stephens RI, Fatemi A, Stephens RR, Fuchs HO. Metal fatigue in engineering. John Wiley & Sons; 2000.
12. Military Handbook SF. Metallic Materials and Elements for Aerospace Vehicle Structures. Washington DC: Department of Defense; 1972.
13. Miner MA. Cumulative damage in fatigue. J Applied Mech 12 1945;12(3):A159–A164.
14. Nanu AS, Marinescu NI, Ghiculescu D. Study on ultrasonic stepped horn geometry design and FEM simulation. Nonconv Technol Rev 2011;4:25–30.
15. Belford JD. The stepped horn: Technical Publication TP-214. 2011.
16. Chen KK, Zhang YS, Chen KK, Zhang YS. Numerical analysis of temperature distribution during ultrasonic welding process for dissimilar automotive alloys. Sci Tech Weld Joi 2015. doi:10.1179/1362171815Y.0000000022.
17. Chen KK, Zhang YS, Wang HZ, Chen KK, Zhang YS, Wang HZ. Study of plastic deformation and interface friction process for ultrasonic welding. Sci Tech Weld Joi 2017. doi: 10.1080/13621718.2016.1218601.
18. Chen K, Zhang Y. Mechanical analysis of ultrasonic welding considering knurl pattern of sonotrode tip. Mater Des 2015;87:393–404. doi: 10.1016/j.matdes.2015.08.042.
19. Shen N, Samanta A, Ding H, Cai WW. Simulating microstructure evolution of battery tabs during ultrasonic welding. J Manuf Process 2016;23:306–14. doi: 10.1016/j.jmapro.2016.04.005.
20. Komiyama K, Sasaki T, Watanabe Y. Effect of tool edge geometry in ultrasonic welding. J Mater Process Technol 2016;229:714–21.

4 Ultrasonic Spot Welding Configuration and Process Parameter Selection

4.1 WELD COUPON CONFIGURATION

A primary objective of various sectors, such as automotive, aircraft, railway transportation, medical, and microelectronics, etc., is to reduce the weight and energy consumption of their products by introducing new and innovative manufacturing techniques. To attain these goals, lightweight and high-strength materials such as aluminum, stainless steel, titanium, magnesium, and copper alloys are necessary. The weldability of these materials depends on their properties, such as ultimate tensile strength (UTS), yield strength (YS), elastic modulus, and hardness. The difficulty in USW processes increases with the increase in hardness of the alloys being joined. However, the softer nonferrous metals can be welded effortlessly by the USW process. Aluminum (Al) is this kind of metal, and it is widely utilized for parts in the aerospace industry, for making fuel cell components in batteries, and for connecting semiconductor devices, transistors, and diodes [1, 2]. The temperature-dependent yield strength of Al has been obtained from a standard reference [3] because, in USW, plastic deformation and yielding occur at elevated temperatures. The complex bonding mechanism of the USW process dictates that yield strength is associated with the temperature rise at both the weld interface and the sonotrode-top part interface. Thus, a linear approximation procedure can be used to determine temperature-dependent yield strength and find the best parameters for satisfactory results.

Ultrasonic metal welding depends on the amount of relative motion taking place at the interfaces of materials. The ultrasonic energy produced by the generator is applied to the top workpiece through the horn, and ultimately it propagates to the weld zone where it produces severe plastic deformation. During this propagation, some energy may be lost due to interatomic friction, so it results in dissipation of heat. The most prominent aspect to be noticed here is specimen thickness. When a thinner material is put on the sonotrode side, there is a better chance of achieving a good weld. Increasing thickness of this part requires a larger sonotrode tip area, as well as a higher level of clamping force and weld power. The maximum thickness of the material that can be joined by USW depends on the material properties of and power levels accessible to the welder. Most of the previous literature [4, 5] confirms that the thickness of the upper specimen has a more significant effect on the weld strength than the thickness of the lower specimen, because the top specimen is

more directly subjected to the impact of ultrasonic energy. The lateral dimensions
(i.e., width and breadth) of the part geometry are the other factors to be taken into
account. The intense ultrasonic vibration that creates the weld in the larger parts
can be transmitted from the weld spot to the surrounding zone, resulting in sticking of the parts to the sonotrode and anvil and negatively affecting any previously
created welds. Previous studies make it clear that the length of the weld coupon
significantly affects the weld strength and quality of the joint, because the waves of
vibration from long and wide specimens are reflected back in an anti-phase manner
[6]. The influence of specimen size on weld strength has been verified experimentally by preparing the specimens according to the ASTM standard with a 20-mm
overlap. Figure 4.1(a) illustrates the coupon dimensions along with extension length.
Figure 4.1(b) shows the various dimensions of specimens used for verification purposes. The six samples for each weld joint are produced with a lateral drive ultrasonic welder. The results with standard-deviation bars are shown in Figure 4.2 for
the 0.3-mm aluminum (AA1100) alloy and copper (UNS C10100), with different
power levels and coupon lengths as an illustration.

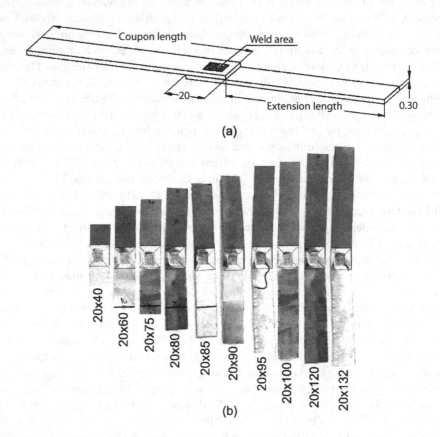

FIGURE 4.1   (a) Schematic diagram showing extension length; (b) lap-welded samples with
different dimensions.

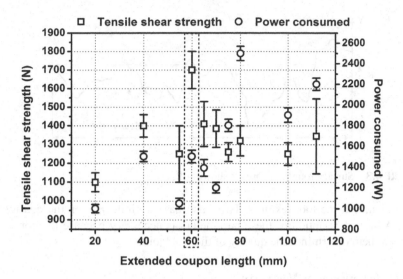

FIGURE 4.2 Effect of extension length on tensile shear strength and power consumed during USW process.

These results are in agreement with DeVries [7], who used a 2.5-kW Sonobond® wedge-reed system to weld 1-mm thick aluminum 6061 T6 alloy. It can be observed from Figure 4.2 that with 20 mm width and 80 mm length (marked in box), good weld strength is achieved with a moderate amount of power consumed. Thus, this is the optimum coupon length, a fact that can also minimize the number of materials that must be used in USW experiments.

The next factor that influences the mechanical strength of the joint produced by the USW process is the material's surface characteristics. It is evident that surface roughness should be considered one of the crucial parameters for heat generation between the surfaces [8]. The substrates on the working surface are measured through the average surface roughness (R_a) term. The roughness of the specimen is measured by taking an arithmetic height of surface irregularities from its mean line over a predefined evaluation length, as shown in Figure 4.3. Eq. (4.1) is used to determine R_a value [9].

$$R_a = \frac{1}{L} \int_0^L |y|\, dx \tag{4.1}$$

4.2 PROCESS PARAMETER SELECTION

There are a number of process parameters that can influence the welding strength, such as frequency of vibration, weld pressure, weld time, weld energy, weld power, and vibrational amplitude. These parameters are regarded as system parameters. Other than these machine parameters, there is another category of parameters called specimen parameters, which include the specimen hardness, sizes, roughness, stack

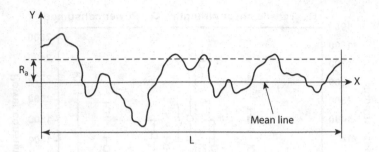

FIGURE 4.3 Surface roughness profile showing R_a.

orientation, and the tool textures. They also have an influence on the quality of a joint. A brief description of these parameters is necessary because the correlation between them determines the quality of the weld joints.

4.2.1 FREQUENCY OF VIBRATION

In ultrasonic systems, the transducer is designed and tuned to operate at a particular frequency. According to the application, the frequency ranges from 15 to 300 kHz. In USW, typically a 20–40-kHz frequency is used for the welding process. The ultrasonic generator is the device that amplifies the normal 50-Hz alternating current supply to 20 kHz of vibration and consistently maintains it during the welding operation. The parts of this system are frequency generator, transducer, power amplifier, booster, and horn. The dimensions of the specimens are also selected in such a way that the whole system will vibrate at a resonant frequency. Notably, a small change in the operating frequency will result in a significant decrease in vibrational amplitude [10]. The frequency may also shift due to changes in temperature, clamping pressure, and/or tool wear. However, the modern integrated power controller with automatic feedback circuitry can compensate if there is any shifting in frequency and make the system resonate at a constant frequency [11].

4.2.2 WELD PRESSURE/CLAMPING PRESSURE/STATIC LOAD

The weld pressure is a key parameter in USW. This pressure exerts a normal force on the specimen throughout the welding process. Thus, in some studies, it is also known as the *static force*. It usually presses the specimens to create intimate contact between the surfaces and is dependent on the weld tip and anvil geometries. The magnitude of this pressure depends on the material's properties and thickness. While adjusting the process parameters, an optimum range of pressure should be defined, because lower weld pressure can lead to a weak weld and higher values may result in excessive deformation of the parts, tip sticking, and generation of high temperatures [12].

4.2.3 VIBRATIONAL AMPLITUDE

The vibrational amplitude is related to the power of the system and the amplification produced by the booster/horn. Thus, it directly reflects the energy delivered to the weld zone. In the ultrasonic welding system, *amplitude* means the axial expansion and contraction of excitation. In a lateral drive system, the amplitude can be controlled by the operator. Originally, this was done by changing the current in the transducer. If the amplitude is set to 100%, then it means that it uses 100% of system power to produce vibrations [13].

4.2.4 Weld Time, Energy, and Power

The quality of the weld in USW also depends on the weld time, which controls the amount of energy and/or power supplied to the joint place. The relationship between the ultrasonic power and weld time varies with the material conditions such as dimension, properties, and surface finish. Figure 4.4 depicts the basic graph between power and time. The area under this curve represents the weld energy (in joules) that is delivered to the weld surface during the weld time. In most USW studies, the weld time and weld energy are often applied conversely to signify the weld power introduced to the weld samples. The weld energy can be represented in the following equation:

$$E = P \times t \tag{4.2}$$

where E represents the weld energy in joules, P is the weld power in watts, and t is the weld time in seconds (sec). If weld power is constant for a particular USW machine, weld energy is directly proportional to weld time. At the "no-loading" condition, the USW system requires a minimum amount of power to maintain the vibration. This power value is directly dependent on the mechanical load [14]. Therefore, an increase in weld pressure and/or mechanical load increases the power level and consequently decreases the weld time to deliver the same amount of energy. Modern USW uses a closed-loop feedback control system to monitor the whole process during welding. In several USW systems, the ultrasonic power, weld energy, and weld time cannot be considered as the independent variable. A better approach to control

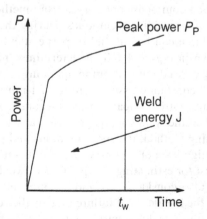

FIGURE 4.4 Weld power variation with weld time [15].

all these parameters is to set a parameter to reach the level of the weld and leave other parameters as the dependent variable(s). For example, if the welding will be done in time-control mode, then a particular weld time will be fixed, and the welding will continue until that time. However, the energy or power that will be achieved during that fixed weld time cannot be controlled. This phenomenon also applies to the energy and power control modes. The power setting in any USW device depends on the high-frequency power entering the transducer, although the actual power at the weld site is different from the power given to the transducer. This depends upon several factors, such as the efficiency of the transducer for conversion of electrical input to mechanical vibration, losses at the interfaces of the transducer-converter-horn assembly, and the dissipation of power from the weld spot to the weld samples and fixture. All these situations make USW a complex process in which different process parameters interact within a very short duration.

4.3   PERFORMANCE MEASURES IN THE USW PROCESS

It is necessary to understand the performance measures that can be obtained from a USW system. Many researchers have tried to get a sound joint between dissimilar materials by controlling the process parameters as well as specimen parameters, such as weld pressure, weld time, weld energy, vibrational amplitude, material hardness, roughness, and material dimensions. These performance measures/outputs of this process can be broadly classified into two categories: (1) mechanical performances, and (2) microstructural characterization. The mechanical performances exhibit the quantitative values of the weld joint, whereas microstructural characterization indicates the quality of the weld. The following mechanical performances can be studied during experiments.

4.3.1   Tensile Shear Load (TS)/Failure Load/Lap Shear Failure Load

The tensile shear load includes the breakage of weld that takes place between the overlapped sheets. It is a common laboratory testing method usually applied to adhesive bonding and other welding techniques. During the test, the direction of application of tensile force can be parallel or perpendicular to the plane of the weld coupons, as shown in Figure 4.5. This determines the failure load results. These tests are usually carried out at room temperature using a universal testing machine (UTM) with a constant crosshead speed of 1 mm/min or 2 mm/min, to avoid any type of dynamic stress impact on the specimen. Meanwhile, the shims are placed at each end of the weld coupons to avoid any type of bending during the test. With this testing method, the shear modulus and strength of a joint can be determined with a high level of accuracy, making it appropriate for the design engineers to use the test for estimating the qualities of weld assemblies. The tensile shear strength is also considered as another kind of response that can be calculated by dividing the tensile shear failure load by the sonotrode tip area. The unit for tensile shear failure load is usually taken as "N" whereas the "MPa" unit is used for tensile shear strength.

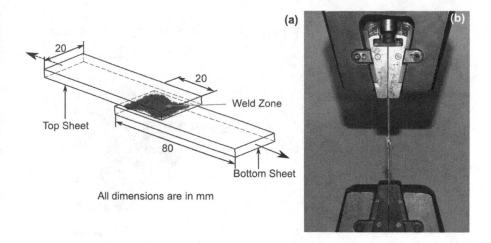

FIGURE 4.5   (a) Tensile shear test coupon; (b) fixture for tensile shear test [16].

4.3.2   T-Peel Failure Load (TP)

The T-peel failure load testing method is commonly utilized to find the peel strength of adhesive joints. However, it is especially valuable to other spot welding processes applied in automotive industries. To assess the potency of the ultrasonically welded joints from an industrial perspective, a regulated approach is necessitated in addition to the existing destructive methods. This method involves the pulling of the specimen perpendicular to the weld surface, as shown in Figure 4.6. Normally, the breaking-force values measured in this test are 20–40% of the tensile shear strength [17]. This method provides information about the peel failure mode during the destructive testing of vehicles by measuring the ultrasonically welded nugget dimensions.

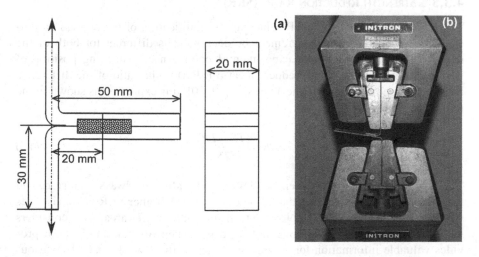

FIGURE 4.6   (a) T-peel test coupon; (b) fixture for T-peel test.

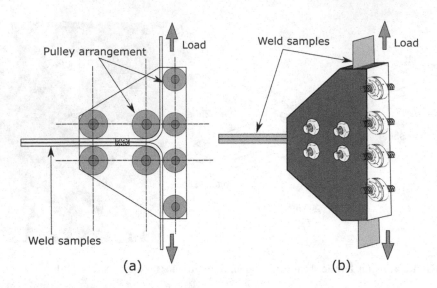

FIGURE 4.7   (a) Schematic drawing of T-peel test fixture, (b) 3D representation of T-peel fixture during testing.

However, the weld nugget formed beneath the footprint of the sonotrode is difficult to find out due to unavailability of a definite weld shape. Thus, a Ford research laboratory (FRL) designed a T-peel fixture that was suitable for testing of long weld strips and maintaining steady sample position during the test (Figure 4.7) [18]. Similar to the tensile shear strength, the T-peel strength can be obtained by T-peel failure load with the sonotrode tip area. The units for T-peel failure load and T-peel strength are same as for the tensile shear failure load and tensile shear strength.

4.3.3   STRENGTH REDUCTION RATIO (SRR)

There is a change in TS and TP values upon modification of the process parameters. However, the rate of increment or decrement is different for both of the results. Thus, for a particular experimental condition and stacking position of the weld metals, the strength reduction ratio (SRR) is the ratio of the difference between TS and TP values to the TS result [19, 20]. It is expressed as shown in the following equation:

$$SRR = \left| \frac{TS - TP}{TS} \right| \tag{4.3}$$

The positive correlation between the TS and TP values is always preferred, and the corresponding SRR value should be close to 1. A higher SRR value signifies stronger joints. This SRR is not reliant on the weld nugget area, as it considers identical experimental conditions and weld specimen orientation. Thus, it provides valuable information for assessing the strength of welds made by various joining processes.

4.3.4 STRESS INTENSITY FACTOR (SIF)

When a crack happens at extreme loading conditions, the thrusting force for the unstable progress of this crack can be measured by the stress intensity factor (SIF). The SIF is produced beneath the sonotrode knurls and inline to the direction of the tensile strength (TS) test. Thus, an equivalent SIF can be obtained from these mixed loading conditions [21].

$$K_{Eq} = 0.694 \frac{F_t}{D\sqrt{t}} \qquad (4.4)$$

where F_t is the peak TS load, D is the nugget diameter, and t is the thickness of the top weld sample. Because most USW machines use a rectangular tip, the area (A) of this tip should be converted to the equivalent nugget diameter (D) by using Eq. (4.5) [22]. The unit of SIF is MPa/m^2.

$$D = \sqrt{\frac{4A}{\Pi}} \qquad (4.5)$$

4.3.5 JOINT EFFICIENCY (JE)

Two types of failure modes—pull-out and interfacial modes—are obtained in the USW process, but the nominal peak stress values for these two modes are different. The tensile stress acquired in pull-out mode can be calculated by the following equation:

$$\sigma_{PF} = \frac{TS_{Peak}}{2\Pi rt} \qquad (4.6)$$

where σ_{PF} is the nominal stress for pull-out mode, r is the weld nugget radius, and t is the thickness of the top sheet. Likewise, the nominal shear stress for interfacial failure mode (τ_{IF}) can be calculated by the following equation:

$$\tau_{IF} = \frac{TS_{Peak}}{\Pi r^2} \qquad (4.7)$$

The joint efficiency (E_w) can be computed by the ratio of the nominal stress values to the ultimate tensile stress (UTS) of the top sheet [23, 24], using the following equation:

$$E_W = \frac{\sigma_W}{S_F} \times 100 \qquad (4.8)$$

where σ_W can be σ_{PF} or τ_{IF} depending upon the failure mode.

4.3.6 TOTAL FAILURE ENERGY/FRACTURE ENERGY (TFE)

During the tensile shear load testing in a UTM, a displacement-load curve is established. From this graph, the total failure energy (TFE) can be evaluated, using the

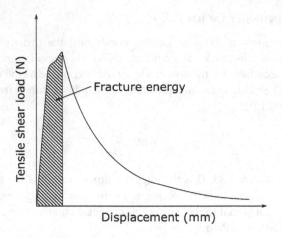

FIGURE 4.8 Schematic illustration of fracture energy.

area under this curve up to the peak load. The TFE has a stronger correlation with the fracture modes than the peak load in tensile shear testing [25]. Thus, the TFE associated with nugget pull-out is higher than the interfacial failure. Figure 4.8 shows a sample TFE. The unit for TFE is "N.mm" or "joule (J)".

4.3.7 U-Tensile Load (UT)

The mechanical strength of a joint can also be expressed in U-tensile load (UT) [26]. A schematic diagram of the test coupon with detailed dimensions is provided in Figure 4.9. The test procedure is same as for the tensile shear test, and the peak load represents the U-tensile failure load. The unit for UT is "N".

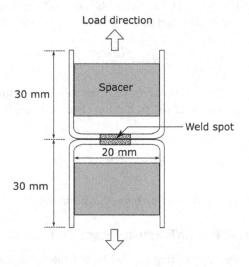

FIGURE 4.9 Schematic diagram of U-tensile load coupon.

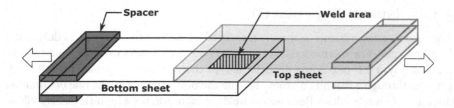

FIGURE 4.10 Schematic diagram of a typical fatigue test specimen.

4.3.8 Fatigue Load/Strength

The structural and automotive applications of welded joints inevitably involve dynamic loading conditions. Thus, fatigue failure of materials and structures is the major cause of accidents. The fatigue tests are performed using a UTM at room temperature with load control mode at various maximum loads. During the tests, the load ratio (R) (i.e., P_{min}/P_{max}) of 0.2 is applied to the weld samples sinusoidally with a frequency of 50 Hz. The maximum load (P_{max}) is considered according to the ultimate TS failure load [27]. The spacers/restraining shims are provided at both ends of weld sample in order to avoid any type of dynamic effect on crack formation. The schematic fatigue test specimen is presented in Figure 4.10.

4.3.9 Weld Area (WA)

Due to the presence of serrations on the horn and anvil surfaces, several point/ line contacts are made between the sheets. At these points, small microbonds initially develop, and when input parameter values are increased, the microbond areas are also gradually increased and are finally saturated. So, it is tough to measure this kind of area, which is known as the "real weld area". Meanwhile, an average or general weld area is calculated by the multiplication of one microbond area with the number of points present on the horn tip (Figure 4.11). The unit of WA is mm^2.

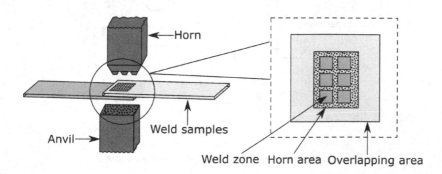

FIGURE 4.11 Schematic illustration of real weld area and average weld area.

4.3.10 INTERFACE TEMPERATURE (T)

The interface temperature is usually measured by K-type thermocouples during the USW process. The thermocouple can be inserted in two ways near the weld spot. In the first way, a 0.1- or 0.2-mm diameter K-type thermocouple is placed near the nugget zone through a machined groove hole on the bottom sheet, in the line of vibration direction (Figure 4.12(a)). Because the welding incorporates a high-frequency vibration and rubbing action between the samples, it is very difficult to confirm that the bead of the thermocouple has remained perfectly at the exact position. At the same time, there may be a chance that the thermocouple may separate the welded samples. Thus, it may slightly affect the temperature measurement. Secondly, to overcome this problem, several thermocouples are attached to the weld interfaces, as shown in Figure 4.12(b). The signal of the thermocouples is recorded by a computer via a data acquisition system at a rate of 25 kHz. The deciding factors for determining the temperature are ultrasonic power, weld time, material properties, thermal conductivity, and geometry of the workpiece. The temperature observed in the weld zone is normally 0.3–0.9 times the temperature of the material having the lowest melting point.

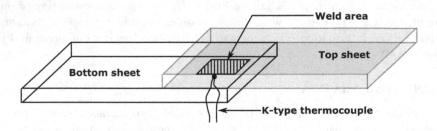

(a) Thermocouple attached to bottom sheet

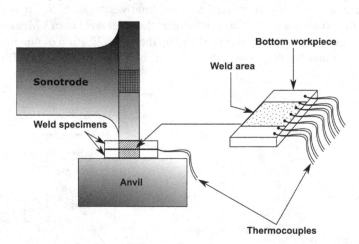

(b) Several thermocouples attached to bottom sheet

FIGURE 4.12   Two types of thermocouple arrangements during USW.

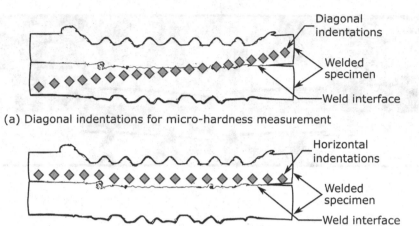

(a) Diagonal indentations for micro-hardness measurement

(b) Diagonal indentations for micro-hardness measurement

FIGURE 4.13 Two types of Vickers micro-hardness measurement.

4.3.11 MICRO-HARDNESS

Hardness is one of the vital factors in the USW process. Hardness is defined as the resistance that the material exhibits to localized plastic deformation. Micro-hardness measurement is the easiest testing method by which to measure the hardness of a material. Thus, the Vickers micro-hardness test is usually performed on a polished weld cross-section using a micro-hardness tester. In particular, these measurements are made at the center of the weld nugget, as well as around the weld spot, to demonstrate the brittleness of the weld. There are two ways to measure micro-hardness: (1) diagonal micro-hardness measurement, and (2) micro-hardness measurement at the weld cross-section (Figure 4.13). The diamond-type indentation made on the specimen during the test is done according to the applied load of 100 g/200 g with a dwell time of 15 sec. These indentations are appropriately spaced to prevent any type of possible strain field effect instigated by contiguous indentations.

4.3.12 ELECTRICAL/JOINT RESISTANCE

The quality of joints produced in the USW process determines the efficiency and lifelong performance of a battery. Unbonded interfaces, cracks, and defects of the joints may cause an increase in resistance and unnecessary energy loss due to the joule heating phenomenon. The electrical resistance of an ultrasonically welded joint can be measured in two ways: (1) using a resistance meter and/ or (2) by applying the four-probe method. Commercially available resistance meters can be employed for the measurement of resistance of weldments according to the ASTM D257-07 standard at room temperature. The span of the resistivity measurements is studied along 1 cm of length [28]. The arrangements for the four-probe method is displayed in Figure 4.14. The two ends of the weld

(a)

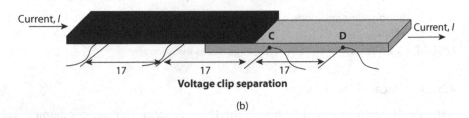

(b)

FIGURE 4.14 (a) Set-up of electrical joint resistance measurement with copper blocks; (b) schematic presentation of positions of voltage clips [29].

specimen are gripped with terminal blocks made of copper or brass with multiple voltage clips, as presented in Figure 4.14(a). The separation between the clips is about 17 mm (Figure 4.14(b)). Figure 4.15 shows a schematic diagram of an electrical circuit which is equivalent to the electric resistance entailed in an ultrasonically welded joint. From this diagram, the formula for overall resistance across the joint is calculated [29].

Resistivity formula:

$$R_{Cu} = \rho \frac{l}{A} \tag{4.9}$$

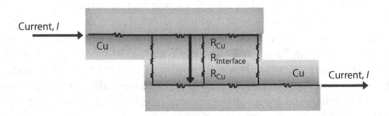

FIGURE 4.15 Schematic illustration of equivalent electrical circuit to demonstrate joint resistance [29].

Joint resistance:

$$R_j = 2R_{Cu} + R_{interface} \qquad (4.10)$$

where R_{Cu} is the resistance (Cu), ρ is the resistivity, l is the length, A is the cross-sectional area and R_j is the overall resistance across the joint.

REFERENCES

1. Ahmed N. New developments in advanced welding. 1st ed. England: CRC Press; 2005.
2. O'Brien RL. Jefferson's welding encyclopedia. American Welding Society; 1997.
3. Kaufman JG. Properties of aluminum alloys: tensile, creep, and fatigue data at high and low temperatures. ASM International. 1999.
4. Park DS, Jang HS, Park WY. Tensile strength of Cu sheets welded by ultrasonic metal welding. Adv Mater Res. 2013:658:202–8.
5. Al-Sarraf Z, Lucas M. A study of weld quality in ultrasonic spot welding of similar and dissimilar metals. J Phys Conf Ser 2012:382:12013.
6. Kodama M. Ultrasonic welding of non-ferrous metals. Weld Int 1989;3:853–60.
7. De Vries E. Mechanics and mechanisms of ultrasonic metal welding. The Ohio State University. 2004.
8. Watanabe T, Yanagisawa A, Konuma S, Yoneda A, Ohashi O. Ultrasonic welding of Al-Cu and AI-SUS304. Study of ultrasonic welding of dissimilar metals (1st report). Weld Int 1999;13:875–86.
9. Zeiss. Handysurf : The Small, Portable Surface Measuring Unit. n.d. https://mediaserver.probes.zeiss.com
10. Graff K. Ultrasonic metal welding, In: Ahmed N (ed.). New developments in advanced welding. Woodhead: Cambridge; 2005.
11. Harman G, Albers J. The ultrasonic welding mechanism applied to aluminium and gold-wire bonding in microelectronics. IEEE Trans Parts, Hybrid Packging 1977;13:406–12.
12. Rozenberg L, Mitskevich A. Ultrasonic welding of metals. vol. 1. Moscow; Plenum Press; 1973.
13. Graff KF, Devine JF, Kelto J, Zhou NY. Ultrasonic welding of metals. vol. 3. Elsevier Ltd; 2007. doi:10.1016/B978-1-78242-028-6.00011-9.
14. What is an ultrasonic weld? n.d. http://jascoes.com/datasheets/WhatIsAnUltrasonic Weld.pdf (accessed August 2, 2014).
15. Matheny MP, Graff KF. Ultrasonic welding of metals. In: Gallego-Juarez J, Graff K (eds). Power Ultrasonics, UK: Elsevier; 2015, 259–93.
16. Satpathy MP, Moharana BR, Dewangan S, Sahoo SK. Modeling and optimization of ultrasonic metal welding on dissimilar sheets using fuzzy based genetic algorithm approach. Eng Sci Technol an Int J 2015;18:634–47.
17. Harthoorn JL. Ultrasonic metal welding (thesis). Eindhoven, Tech Hogeschool, Dr Tech Wet Diss 1978 150 P. 1978.
18. Ward S, Hetrick E, Jahn R, Reatherford L, Grima T, Wilkosz D. Part II: Ultrasonic metal welding enabling the aluminum vehicle. 2005. https:// files.aws.org
19. Das A, Li D, Williams D, Greenwood D. Weldability and shear strength feasibility study for automotive electric vehicle battery tab interconnects. J Brazilian Soc Mech Sci Eng 2019;41: 54–68. doi:10.1007/s40430-018-1542-5.
20. Han L, Thornton M, Shergold M. A comparison of the mechanical behaviour of self-piercing riveted and resistance spot welded aluminium sheets for the automotive industry. Mater Des 2010;31:1457–67.

21. Zhang S. Stress intensities at spot welds. Int J Fract 1997;88:167–85.
22. Mohammed SMAK, Dash SS, Jiang XQ, Li DY, Chen DL. Ultrasonic spot welding of 5182 aluminum alloy: evolution of microstructure and mechanical properties. Mater Sci Eng A 2019;756:417–29. doi:10.1016/j.msea.2019.04.059.
23. Lu Y, Mayton E, Song H, Kimchi M, Zhang W. Dissimilar metal joining of aluminum to steel by ultrasonic plus resistance spot welding: microstructure and mechanical properties. Mater Des 2019;165:107585. doi:10.1016/j.matdes.2019.107585.
24. Bohr J. A comparative study of joint efficiency for advanced high strength steel. USA. 2009. www.autosteel.org
25. Bakavos D, Prangnell PB. Mechanisms of joint and microstructure formation in high power ultrasonic spot welding 6111 aluminium automotive sheet. Mater Sci Eng A 2010;527:6320–34.
26. Lee SS, Kim TH, Hu SJ, Cai WW, Li J, Abell JA. Characterization of joint quality in ultrasonic welding of battery tabs. ASME 2012 Int Manuf Sci Eng Conf Collocated with 40th North Am Manuf Res Conf Particip with Int Conf, MSEC 2012. 2012;135:249–61. doi:10.1115/MSEC2012-7410.
27. Patel VK, Bhole SD, Chen DL. Fatigue life estimation of ultrasonic spot welded Mg alloy joints. Mater Des 2014;62:124–32.
28. Mohan Raj N, Kumaraswamidhas LA, Nalajam PK, Arungalai Vendan S. Studies on electro mechanical aspects in ultrasonically welded Al/Cu joints. Trans Indian Inst Met 2018;71:107–16. doi:10.1007/s12666-017-1140-8.
29. Shin HS, de Leon M. Mechanical performance and electrical resistance of ultrasonic welded multiple Cu-Al layers. J Mater Process Technol 2017;241:141–53. doi:10.1016/j.jmatprotec.2016.11.004.

5 Ultrasonic Spot Welding of Dissimilar Metal Sheets

Generally, welding of dissimilar metals is warranted in two cases. The first is when a combination of two metals is required over a span of length to achieve complementary properties. The second is when the service condition on either side of the welding point is suitable for different environments and suits two different materials. This is often a complicated process owing to the fact that the two materials may not be a natural fit for each other. The welding parameters suitable for one material may not suit the other. Hence, the differing material properties, such as melting point, density, thermal and electrical conductivity, crystal structure, mutual diffusibility, and ductility, among others, affect the joining process. Despite all these difficulties, dissimilar welding has been increasingly adopted for a multitude of applications, though it is not limited to the ones described here.

5.1 ULTRASONIC SPOT WELDING OF ALUMINUM TO COPPER

Lithium-ion (Li-ion) batteries are now extensively used in many automotive manufacturing industries due to their high energy storing capacity, reliability, safety, robustness, and lightweight nature [1]. Most of the Li-ion batteries used in automobiles are in pouch cell format. This format employs tab-to-busbar interconnects for the transmission of electrical power. The combination of these pouch cells forms a module, and hundreds of modules are interconnected in a battery pack that ultimately defines the total power of the Li-ion battery pack.

The busbar is the primary element that determines the effectiveness of Li-ion batteries, as it provides the desired electrical, thermal, and mechanical properties. The energy-carrying capacity and the cost of Li-ion batteries are mainly based on the selection of busbar material and thickness [2]. The materials used in the busbar should have high electrical and thermal conductivities and be of suitable thickness to avoid excessive heat generation because of the joint resistance. Typically, aluminum (Al) and copper (Cu) are commonly and extensively used materials, not only in the domains of busbars, power device module packing, and lithium-ion batteries assembly, but also in microelectronic technology [3]. Unfortunately, the joining of these materials by a fusion spot-welding process produces bulk and brittle intermetallic compounds (IMCs), a high level of weld distortion, and merely average joint strength [4]. For these reasons, an alternative solid-state welding process was developed that did not have these inferior properties. Ultrasonic spot welding (USW) is an environmentally friendly and low-cost method for producing a joint between dissimilar sheets within a few seconds.

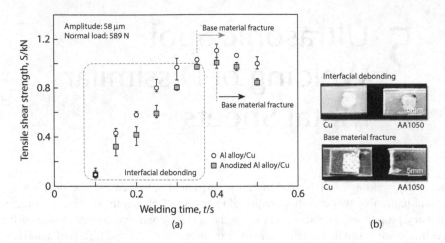

FIGURE 5.1 (a) Comparison of tensile shear strength of AA1050/Cu welded samples at various weld times; (b) samples fractured during lap shear strength tests [5].

The heat produced at the weld zone by vibration amplitude, weld pressure, and weld time influences the weld integrity in terms of tensile shear and T-peel strengths. Figure 5.1(a) demonstrates the relationship of lap shear tensile strength to various welding times of the anodized and non-anodized AA1050/Cu samples. It is clearly ascertainable that the samples are not appropriately welded when the welding time is less than 0.35 sec. The lap shear tensile strength increases gradually as the welding time increases. Figure 5.1(b) presents a traditional lap shear tensile tested sample that has crack surfaces with two types of fracture modes. The fracture modes change from interfacial debonding to base material fracture when the welding time increases to a particular level during the USW process. For base-material fractured samples, the lap shear tensile strength increases only up to a certain limit; after that it decreases further with increasing welding time. This is due to the vigorous softening of aluminum alloy and the higher amount of plastic deformation that occurs in the weld zone at the elevated temperature. At various welding times, lap shear strength of the non-anodized aluminum-based samples is higher than that of anodized aluminum-based samples. This is due to the presence of residual anodic aluminum oxide layer on the weld zone, which directly affects the mechanical strength of Al/Cu samples.

It is well understood that the surface conditions influence the weld quality and strength of the joint. The relationship between tensile shear and T-peel failure loads with the various weld times is described for four different surface conditions in Figure 5.2. Firstly, for the lubricating surface condition, ethanol is utilized on the faying surface. Initially, these two failure loads remain relatively constant up to a certain weld time, followed by a rapid increment of these failure loads. Thus, it is shown that adhesive wear takes place because of the evaporation of the ethanol at the high welding time and vibration amplitude. This creates an appropriate atmosphere for the plastic deformation of the Al sheet. Next, after reaching the maximum limit of control factors, the failure loads decrease suddenly because of extreme weld time, and cracks can be spotted at the edge of the weld region. Secondly, for the normal faying

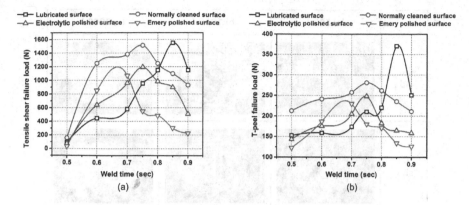

FIGURE 5.2 Tensile shear and T-peel failure loads of Al (AA1100)-Cu (UNS C10100) weld samples for different surface conditions: (a) tensile shear failure load; (b) T-peel failure load [6].

surface condition, both failure loads show maximum shear strength up to a certain limit. After that, shear strength decreases gradually with the weld time due to the crack formation that occurs in the weld area. Similarly, in the case of electrolytic polished and emery polished surface conditions, initially there is a high-pitched increase in these failure loads. However, the values of these failure loads are less than that of the normal state, but more than the lubricating surfaces. The condition that leads to this phenomenon is that at the starting stage of welding, the electrolytic and emery polished surface roughness value is higher compared to the rest of the surfaces, and it causes high-temperature generation with excessive material softening.

The microhardness variation at the welding and adjacent heat-affected zone demonstrates the occurrence of plastic deformation in the aluminum sheets during the USW process. Figure 5.3 reveals the microhardness variation of the weld region at different welding conditions. The minus side indicates the distance of the hardness distribution of Al from the weld region, and the plus side indicates the distance of

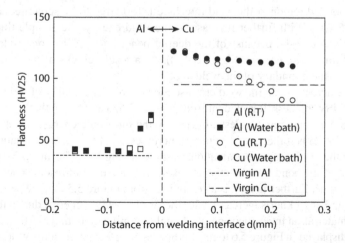

FIGURE 5.3 Microhardness variations at different welding conditions [7].

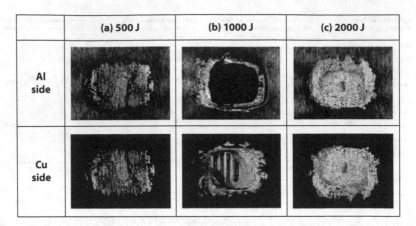

FIGURE 5.4 Fracture surfaces of ultrasonically welded Al (AA 6061)-pure Cu joint obtained with various welding energies [4].

hardness distribution of the Cu side. The increasing tendency toward hardness in the area of the weld in both atmospheric and underwater welding conditions indicates finer crystals and strain hardening in the weld region. In the case of atmospheric welding, the welded region temperature increases gradually due to the recrystallization of material by the annealing effect; thus, the hardness is lowered. However, in the case of underwater welding, a comparative high microhardness value is obtained due to the suppression of heat at the welded area.

The specific reason for the tensile shear and T-peel failure load variations can be perceived from a fracture surface analysis (Figure 5.4). An evolution of failure modes can be clearly detected with the increase in welding energy. At lower welding energy, the interfacial detachment at the welded area is clearly visible, with micro-welds on both the Al and Cu sides (Figure 5.4(a)). When the welding energy increases, the size and number of micro-welds are also increased. The weld pull-out failure arises when the bonded region at the weld interface is too large to be separated, as shown in Figure 5.4(b). With further increase in the weld energy, the sample thickness is significantly decreased because of the deeper penetration of the sonotrode tip with intensified ultrasonic energy (Figure 5.4(c)). As a result, cracks occur in the aluminum sheet at the boundary of the weld spot.

Optical microscopy on the weld cross-section exposes the quality of weld produced during the USW process. It is evident from Figures 5.5(a) and 5.5(b) that at low welding energy, the bond line remains less affected by the ultrasonic energy, and it is macroscopically flat. When the welding energy increases steadily, a small variation can be detected in the macroscopic examination, as shown in Figures 5.5(c) and 5.5(d). The rate of waviness of the bond line directly relates to the amount of ultrasonic energy passed to the weld zone. Furthermore, it is also obvious from Figure 5.5 that the penetration of sonotrode and anvil knurls increases with the rise in welding energy, due to the softening of the materials at intensified energy. A high-magnification image of the weld cross-section is displayed in Figure 5.6 to reveal the plastic deformation at the weld zone. The weld interfaces at the aluminum side are composed of a wavy or swirl-like pattern.

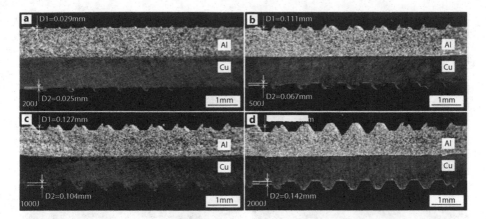

FIGURE 5.5 Optical microscopy of weld cross-section of AA 6061/Cu weld samples at various welding energies [4].

When the welding energy is increased, the penetration of the deformation zone is also increased significantly. A wavy bond line formation mainly enhances the joint strength between the materials. However, when the weld energy surpasses a critical value, the bond line becomes too wavy on the aluminum side. As a result, lots of voids are formed in the welded area. The presence of these voids may act as fracture initiation sites during lap shear tensile testing and decrease the joint strength and ductility.

The SEM analysis of a fractured surface can expose the degree of plastic deformation and microbonds at the weld zone (Figure 5.7). As Al is a comparatively softer material than Cu, more plastic deformation occurs at Al side, as can be seen in Figure 5.7(a) and (c). In these figures, sonotrode and anvil prints on the weld samples are clearly visible. It is expected that the intermetallic bond formation will happen between the two sheets, and it affects the joint strength significantly. Figure 5.7(b) and

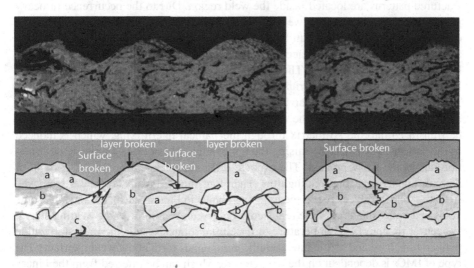

FIGURE 5.6 High-magnification optical microscopy images of Al tab interfaces [8].

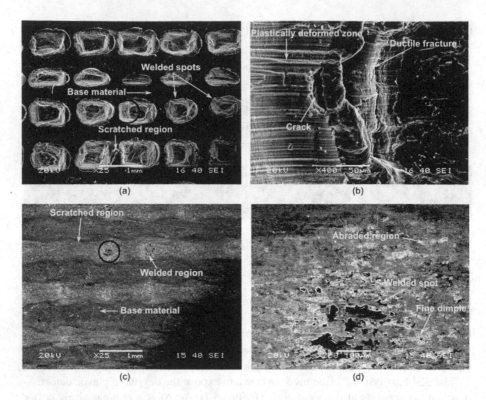

FIGURE 5.7 SEM images of fractured surfaces of ultrasonically welded Al and Cu samples: (a) Al side (b) magnified image of a (c) Cu side (d) magnified image of (c) [6].

(d) are the magnified images of the black-circled area. In these figures, the vertically fractured patterns are located inside the weld region. Due to the occurrence of heavy plastic deformation around the weld region, a crack has also formed on the Al side. The effect of the hardness of Cu can be clearly observed from Figure 5.7(d). Some of the zones of the Cu fractured surface are subjected to wear only, and no bond formation has happened in these zones. The microbonds look like fine dimple-like structures. Figure 5.8 shows SEM micrographs of the weld cross-section at different weld times. Two types of welding region can be noticed in this SEM analysis. In both figures, the oxide layers are cracked into the Al matrix. These oxide layers and Cu elements may be transfused into the Al matrix through mechanical intermixing, which is induced in the softened Al matrix by raising the temperature of interface by plastic deformation during the USW process. The average diffusion thickness of the Al oxide layer increases with increasing welding time. Therefore, most Al oxide layers were removed from the weld interfaces in the specimen, as shown in Figure 5.8(b). Meanwhile, some voids and noticeable cracks are repeatedly observed even in the early welding stages. Whenever the interface temperature of dissimilar joints exceeds the recrystallization temperature, hard and brittle intermetallic compounds form at the weld interface. The type of IMCs is dependent on the temperature, which can be inferred from the binary phase diagrams of the metals. Figure 5.9 displays the Al/Cu phase diagram.

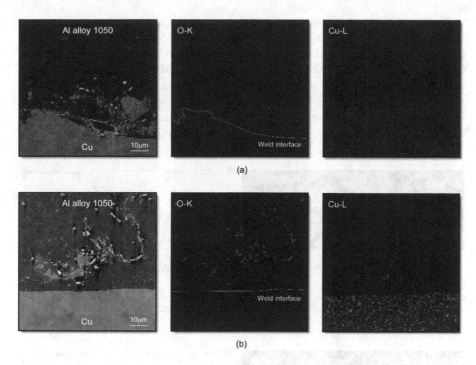

FIGURE 5.8 SEM images of weld interfaces at different weld time: (a) 0.2 sec; (b) 0.4 sec [5].

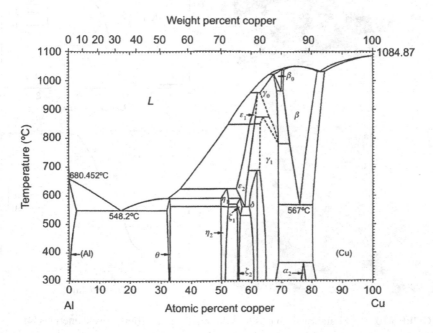

FIGURE 5.9 Al-Cu binary phase diagram [9].

Further, the diffusion of elements and their thickness, analyzed by back-scattered SEM images along with EDS chemical analysis (Figure 5.10), can forecast IMC formation. Figure 5.10(a) illustrates a joint made at lower weld energy. The thickness of this diffusion zone is around 4–7 μm. Meanwhile, the IMC layer can hardly be seen at the weld interface when welding energy is minimum. This result confirms that no continuous IMC layer is formed in the welded region. However, this thickness increases with the rise in weld energy, and it can be perceived from Figure 5.10(b) and (c). Once again, Figure 5.10(c) proves that at maximum weld energy, many

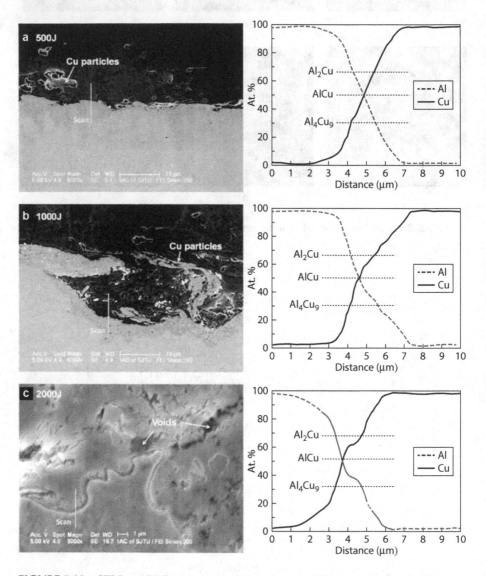

FIGURE 5.10   SEM and EDS images of weld interfaces at different weld energies [4].

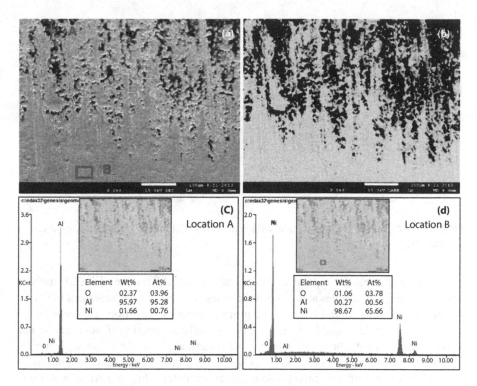

FIGURE 5.11 SEM images of Al side: (a) secondary electron image; (b) back-scattered electron image; (c) and (d) EDS chemical analysis at locations A and B [8].

cracks are formed at the periphery of the weld zone. Numerous IMCs can form in the Al/Cu joint, such as Al Cu, Al_2Cu, and Al_4Cu_9. This type of IMC layer creates detrimental effects on joining strength. Furthermore, the chemical composition on the Al fractured surface is clearly confirmed by the EDS volume scan analysis (Figure 5.11). The contrast of these images is based on sample surface topology as well as atomic weight. In case of a heavier metal, more electrons are reflected with lighter color and stereo surface topology features. The black-scattered SEM image contrasts are more sensitive to the atomic weight, and they shows a better black/white contrast. Thus, the volume scan on the black abraded region reveals the Al element and very little amounts of the Ni element (as the welding is performed between Al and Ni-coated Cu). The white look-like region displays the Ni element only.

XRD analysis is another technique for recognizing the phase of IMCs formed during the joining process. Figure 5.12 shows the traditional fracture surface and XRD forms of the AA 1060/Cu joints. The XRD analysis confirms that the Al_2Cu brittle IMC layer can cause low joint strength along with fracture at the welded surface. The joint strength is found to decrease with further increases in the welding time, due to the generation of more brittle Al_2Cu IMCs.

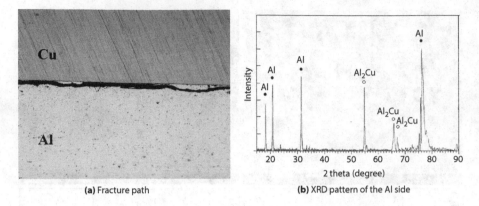

FIGURE 5.12   (a) Weld cross-section with fracture line; (b) XRD analysis on weld cross-section [10].

Figure 5.13 presents typical electron backscatter diffraction (EBSD) analysis results on the weld samples at various weld times. The black lines represent the grain boundaries of both Al and Cu, and the grain orientation directions can be inferred from the color scale legend in the inverse pole to rolling direction of weld sheets. The white arrows indicate the location of the weld interface. In all three cases, no significant amount of change is noticed on the Cu side due to its comparatively higher hardness. However, significant perceptible changes are detected on the Al side. Around the weld region, the Al side microstructure changes significantly with the increase in weld time, due to the incidence of severe plastic deformations during the USW process. The thickness of the deformed weld region varies with the location due to the presence of knurled-type textures on the sonotrode tip. In contrast, the volume

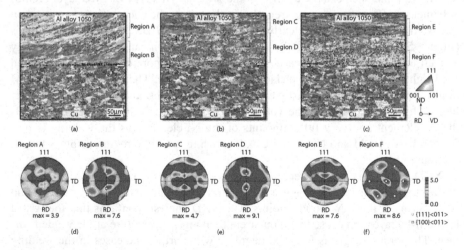

FIGURE 5.13   EBSD images around the weld interface of AA 1050/Cu samples with inverse pole figures at different welding times [5].

FIGURE 5.14 (a) Bright-field and (b) annular dark-field TEM images at the weld interface of Al (AA 1050) and Cu joints [5].

of plastic deformation in the valley region placed under a knurl can be smaller than that in the region located outside of the knurled tip. Indeed, the microstructure in the vicinity of the joint region changes significantly as welding time increases.

Transmission electron microscopy (TEM) analysis can be utilized to comprehend the complete microstructures in the weld region. Figure 5.14(a) and (b) exhibits a bright field and annular dark field at the weld interface between Al (AA1050) and Cu weld samples. The grain size of the aluminum material at the joint region is approximately 100–200 nm (Figure 5.14(a)). Additionally, various sizes of Al material grains are observed in the welded area. These grains are generated during bond formation in the weld region [11]. Some area of the weld interface is also found to be a reactive layer (Figure 5.14(b)). The thickness of these grains is about 40–100 nm. In this figure, an Al_2Cu IMC is detected by the chemical analysis using the TEM-EDS (energy dispersive spectroscopy) process. This thin IMC layer contributes significantly to the enhancement of weld strength in the USW of Al/Cu specimens. However, the increase in the thickness of the IMC layer has an adverse effect on joint strength and quality.

5.2 ULTRASONIC SPOT WELDING OF ALUMINUM TO MAGNESIUM ALLOYS

Aluminum (Al) and magnesium (Mg) alloys are the two commonly preferred materials in the automotive industries due to their low density, high strength-to-weight ratio, and recyclability properties [12, 13]. Thus, the demand to replace steel in the structural elements of an automobile is continuously increasing. However, it is a challenge to produce good-quality Al/Mg joints using the conventional fusion welding process, due to the rapid formation of hard and brittle IMCs such as $Al_{12}Mg_{17}$ and Al_3Mg_2. These IMCs deteriorate the joint strength; thus, control of the formation of these IMCs is inevitably necessary [14–16]. Compared with a fusion welding process, the solid-state welding process occurs at low temperature, and the rapid conversion between solid and liquid phases does not happen in this process. This process suppresses the formation of Mg-Al brittle IMCs, cracks, and pores.

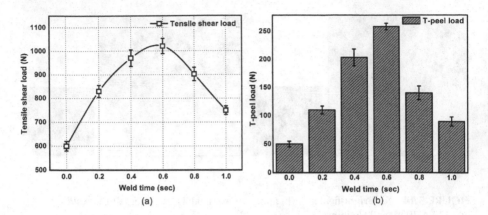

FIGURE 5.15 Variation of tensile-shear and peeling forces of ultrasonically welded AZ31B/AA6061 joints with weld energy.

All the available literature on USW of Al-Mg alloys indicates that the formation of an IMC layer at the weld interface is a common feature irrespective of the use of various welding parameters or stack configurations or alloy systems. The thickness of this IMC layer ultimately determines the weld strength. The tensile-shear and peeling force variations of ultrasonically welded AZ31B/AA6111 joints with various weld energies are presented in Figure 5.15. The tensile shear force is initially increased, and then is decreased with further increase in weld time (Figure 5.15(a)). Meanwhile, peel tests are typically performed to ensure the bonding ability between the weld surfaces. The results of this peel test also exhibit a nature similar to that revealed by the tensile shear force curve (Figure 5.15(b)). At the low weld time values, the majority of the joints fail due to the pull-out of Al alloy weld nuggets. This fracture develops small "fish scale"-like facets indicating inadequate plastic deformation at the weld zone. However, when the weld time is at its peak level, the decrement in the joint strengths is attributable to dislocation fracture, with many secondary cracks at the edge of the weld spot. Moreover, substantial weld energy input also produces a hard and brittle IMC layer at the weld interface. Thus, when a crack is initiated, it will propagate rapidly, and eventually the joint will fail.

Furthermore, the strength of ultrasonically welded joints depends on the pattern of the sonotrode knurls and vibration amplitude. The results of lap shear tests for AZ31B–AA5052 joints with various weld energies and vibration amplitudes under three sonotrode knurl patterns are displayed in Figure 5.16. It is clearly observable that the lap shear load increases considerably when the vibration amplitude rises from 65% to 85% for the A-type sonotrode pattern. An increase in vibration amplitude results in an increment of welding power applied to the weld spot and a reduction of weld time. However, when the vibration amplitude exceeds the critical limit of 85%, the joint strength subsequently decreases due to excessive reduction of sheet metals and the formation of cracks at the weld edge. However, the geometrical form of A-type sonotrode knurls is of the rectangular pyramidal type, which effectively drives the upper sheet against the lower sheet, thus promoting the occurrence of

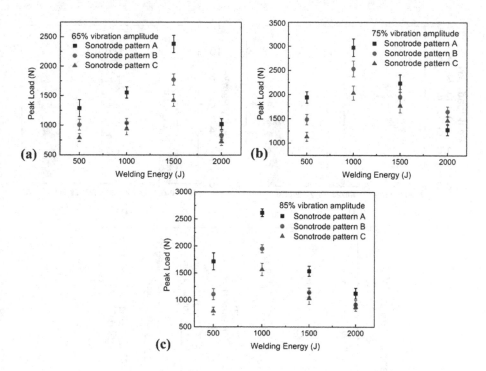

FIGURE 5.16 Peak tensile load of Mg/Al ultrasonically welded joints with various vibration amplitudes and weld energy levels under three sonotrode knurl patterns [17].

plastic deformation between the sheets. This type of sonotrode tooth pattern also prevents excessive thinning of the sheets.

The critical stress intensity factor (K_c) is another way to represent tensile shear strength. This factor is used to normalize the effect of weld energy on the tensile shear strength, and K_c can be compared for various weld combinations. This factor can be computed according to Zhang's formula [18, 19]:

$$K_c = 0.694 \frac{F_t}{d\sqrt{t}} \tag{5.1}$$

where F_t is the ultimate tensile load, t is the thickness of the sheet, and d is the weld nugget diameter. During the welding of dissimilar sheets, the thickness is taken as the average value, i.e., $t = (t_1 + t_2) / 2$, where t_1 and t_2 are the thickness values of two layers respectively. This concept was initially used for the resistance spot welding process. Thus, the d value should be represented in terms of area (A) for the ultrasonic welding process, as it holds a rectangular weld tip. The relationship can be calculated from the following expression:

$$d = \sqrt{4A/\pi} \tag{5.2}$$

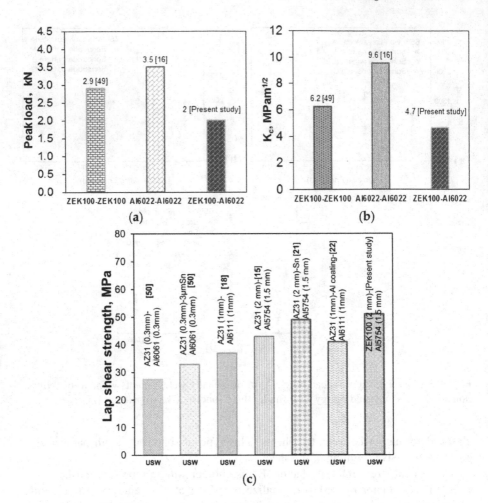

FIGURE 5.17 Comparison of (a) peak load, (b) critical stress intensity, and (c) tensile shear strength results of ZEK100-Al6022 dissimilar joints with results from other literature testing different Al and Mg alloys [18].

Figure 5.17(a) and (b) demonstrates the comparison results of peak tensile load and K_c value of dissimilar ZEK100 Mg alloy-AA6022 Al alloy with similar AA6022-AA6022 and ZEK100-ZEK100 joints. It can be noticed that the peak load, as well as the K_c values for dissimilar ZEK100 Mg alloy-AA6022 Al alloy, are less than the results obtained in similar joints. This is due to the formation of brittle α-Mg+Al$_{12}$Mg$_{17}$ IMCs during the welding of this dissimilar metal combination. Moreover, the peak load and K_c values of AA6022-AA6022 joints are higher than those of ZEK100-ZEK100 joints. This would be associated with differences in physical, thermal, and mechanical properties such as density, formability, thermal conductivity, specific heat, diffusivity, etc. Figure 5.17(c) presents the tensile shear strength results of various Mg and Al grade metals during ultrasonic spot welding. In all the cases, the faying surface roughness and the sonotrode tip area remains the

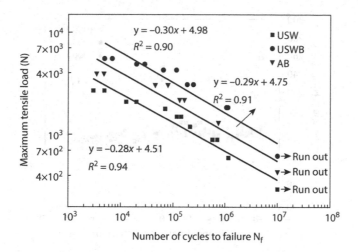

FIGURE 5.18 Comparison of fatigue strengths of Mg/Al joints produced with USW, AB, and USWB methods [21].

same. The lap shear strength of the ZEK100-AA6022 dissimilar metal combination outperforms other metal combinations due to the formation of a low-thickness brittle diffusion layer.

A comparison of the fatigue strength results of AA5052 Al alloy-AZ31B Mg alloy joints produced by USW, adhesive bonding (AB), and ultrasonic spot weld bonding (USWB) is shown in Figure 5.18. This figure reveals the relationship between maximum fatigue load and number of cycles to failure (N_f); if a joint does not undergo fatigue failure even after 10^7 cycles, then this condition is treated as runout. The curves fitted to the fatigue data in these three welding conditions seem to be parallel. The highest tensile fatigue load is obtained for USWB specimens, followed by AB and USW. Because the USWB method adds adhesive to the faying zone, the stress concentration at the weld edges can be effectively relieved, and the weld zone is entirely under uniform stress distribution. The addition of adhesive also restricts the formation of brittle IMCs in the weld cross-section. A similar phenomenon is also observed in friction stir spot welding (FSSW) of Al/Mg alloys [20].

During the welding of dissimilar metals, USW creates no apparent fusion or heat-affected zone (HAZ), which is the primary source for degradation of joint strength. It can be inferred that USW involves a solid-state bonding mechanism, and the lower thermal conductivity of a sonotrode tip as compared to weld metals restricts the HAZ. The microhardness profile of AA5754-AZ31 (Al-Mg) alloy joints across the bonding cross-section is represented in Figure 5.19. It can be easily understood that the microhardness values decrease with the increase in weld energy, and this is attributed to the increase in the grain size at elevated temperatures. The hardness observed exactly at the weld center is about 200–300 HV. The sudden increase in the hardness value is due to the formation of brittle $Al_{12}Mg_{17}$ IMCs. Thus, the microhardness measurement also reveals the presence of brittle IMCs at the weld cross-section, which is the primary cause of the development of weld fractures.

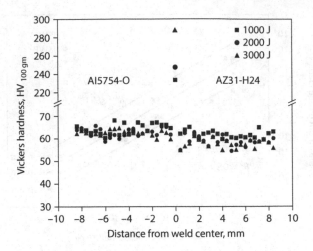

FIGURE 5.19 Microhardness at weld cross-section for various weld energies [22].

A study of the fracture surfaces of Al-Mg alloys provides overall information about different weld qualities at various parametric conditions. Figure 5.20 shows the fractured surfaces of similar AA6111-AA6111 Al alloys, AZ31-AZ31 Mg alloys, and AA6111-AZ31B Al-Mg dissimilar alloys. The weld conditions remained identical for the welding of these three combinations. It may be observed that the AA6111-AA6111 welds impart the maximum weld strength, followed by the AZ31-AZ31 and AA6111-AZ31B alloy combinations. However, the nature of the weld strength profile with respect to the weld energy is similar for all three conditions. At lower energy, the failure occurs at the interface due to the low density of microbonds, and when the weld energy rises, the failure mode changes to nugget pull-out. Figure 5.20(a) and (b) depicts the nugget pull-out failure mode at the higher weld energy values. One can see that the fracture energy of AZ31-AZ31 similar Mg alloys is about half that of the AA6111-AA6111 Al alloys. The dissimilar AA6111-AZ31B welds display a rapid decrement in lap shear strength at elevated weld energy or longer weld time. Moreover, the interfacial failure mode is always noticed at higher weld energies (Figure 5.20(c) and (d)).

An electron probe micro analysis (EPMA) image on the fractured bottom surfaces (on the anvil side) of both AZ31B and AA5052 sheets is shown in Figure 5.21. White scratch marks indicate the welded portions, and the microbonds grow only in these zones. Outside this zone, the black region represents the scratched zone, where only negligible friction occurs. The results demonstrate that when the Al is on the sonotrode side and the Mg is on the anvil side (Figure 5.21(a)), more scratched portions are obtained showing the unbonded/undeformed zones. The intensity of Al is also much less traced on the Mg side. In contrast, when the Mg is on the top (Figure 5.21(b)), the deformation zone is clearly visible, and it is almost equal to the diameter of the sonotrode tip. Meanwhile, a higher concentration of Mg is traced on the Al side, revealing the effective atomic interaction of Mg with Al.

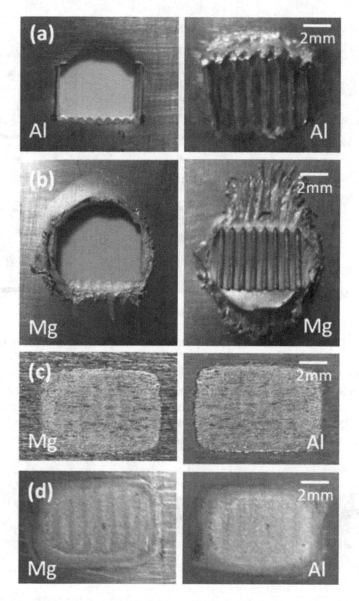

FIGURE 5.20 Overview of fractured surfaces of (a) Al-Al weld, (b) Mg-Mg weld, (c) Mg-Al weld at low energy and (d) Mg-Al weld at high energy [23].

Optical microscopy images of the AA5052-AZ31B weld cross-sections at 70% of vibration amplitude and 0.6 sec of weld time are presented in Figure 5.22. Figure 5.22(a) shows the weld cross-section when the Al is on the sonotrode side, whereas Figure 5.22(b) presents a weld cross-section with Mg on the sonotrode side. A prominent indentation mark is observed on the specimen when Mg is on the top side. This is attributed to the low hardness of AZ31B Mg sheets as compared to

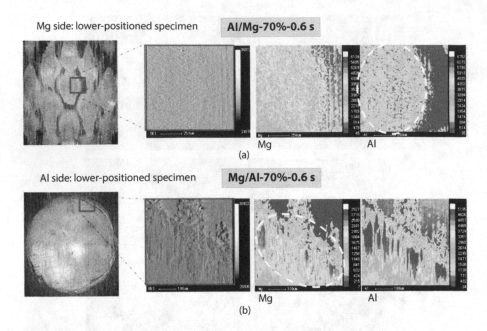

FIGURE 5.21   EPMA images of fractured surfaces of (a) Al/Mg and (b) Mg/Al weld samples [24].

the AA5052 Al sheet. Magnified sections of the cross-sections reveal the possibility of formation of an IMC layer due to interatomic diffusion. A thicker IMC layer (~10 μm) formed when the Al was placed on the top, and a comparatively lower thickness IMC layer (~5 μm) formed when the Mg was placed on the horn side. This IMC layer formed due to the occurrence of a eutectic reaction between the Mg

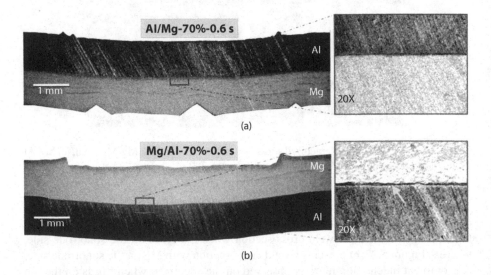

FIGURE 5.22   Optical microscopy of (a) Al/Mg and (b) Mg/Al weld cross-sections [24].

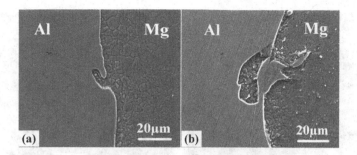

FIGURE 5.23 Mechanical interlocking feature of Al/Mg ultrasonic spot welds at different weld energies [17].

and Al specimens during the USW process. The presence of these brittle interlayer phases determines the joint strength, and basically their presence depends on the lap configuration of the metals.

The morphology of interfaces, microstructure evolution, and bonding mechanisms during USW of Al-Mg alloys under various parametric conditions can be analyzed with electron microscopy. The bonding mechanism during high-power ultrasonic welding of AA5052 Al alloy-AZ31B Mg alloy under different welding energies is presented in Figure 5.23. As the pyramidal teeth on the sonotrode knurls sink into the top sheet, the severity of plastic deformation and thinning of the specimen depend upon the level of ultrasonic energy passed to the faying surface. It can be seen that no noticeable IMCs are formed at the weld interface during a short weld energy cycle. However, the thickness of the IMC formation increases with the rise in weld energy values. The vortex-like region in Figure 5.23(a) shows the effective bonding at the interface, and this region is considered a mechanical interlocking zone. Nevertheless, this vortex-like region is enlarged with the increase in weld energy (Figure 5.23(b)). This non-uniform and complex material flow causes effective plastic deformation at the interface, which hinders the spreading of the intermetallic layer throughout the faying surface.

Fractrographic SEM images of ultrasonically spot welded AA6022-ZEK100 dissimilar metals are presented in Figure 5.24. As already stated, the interfacial failure mode is always obtained for all levels of weld energy, and in the case of similar Al/Al and Mg/Mg joining, the failure mode changes from interfacial to nugget pull-out. This is attributable to the different bonding mechanisms and flow behavior in joining of similar and dissimilar metals. The high-stress region around the weld edge is present on both the sides (i.e., the Al side and the Mg side; Figure 5.24(a) and (b)). The yellow dotted boxes on Figure 5.24(a) and (b) are magnified in Figure 5.24(c) and (d). The EDS volume scan of the Al fractured surface reveals the maximum constituents, such as 59.3 at. % Al and 39.9 at. % Mg. Likewise, the fractured surface of the Mg side consists of 63.8 at. % Mg and 35.8 at. % Al. The fact that both the surfaces contain Al and Mg elements confirms the occurrence of interface diffusion across the bond line, and it may be composed of an α-Mg+Al$_{12}$Mg$_{17}$ IMC. Once again, the yellow dotted boxes in Figure 5.24(c) and (d) are magnified in Figure 5.24(e) and (f). The EDS point scan on the black region of Figure 5.24(e) represents the major

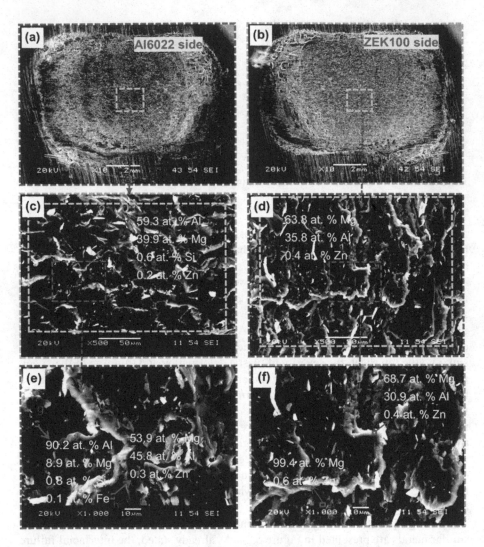

FIGURE 5.24   SEM images of fractured surfaces of ultrasonically welded AA6022/ZEK100 dissimilar joints: (a) Al side, (b) Mg side, (c, d) lower magnification of both surfaces and (e, f) higher magnification of both surfaces [18].

composition of 90.2 at. % Al and 8.9 at. % Mg. Similarly, the EDS analysis on the white spotted region reveals 53.9 at. % Mg and 45.8 at. % Al. However, the black region of Figure 5.24(f) contains 99.4 at. % Mg and 0.6 at. % Zn elements. The white spot of Figure 5.24(f) consists of 68.7 at. % Mg and 30.9 at. % Al. It indicates the non-uniform distribution of an α-Mg+Al$_{12}$Mg$_{17}$ IMC across the interface.

The static diffusion study reveals that when the Al and Mg weld samples are not suitably deformed under the influence of USW process parameters, two types of continuous sublayers (such as Mg$_{17}$Al$_{12}$ and Al$_3$Mg$_2$) are developed on the Mg and Al sides respectively, due to intermetallic reaction. The kinetics of the growth of the IMC

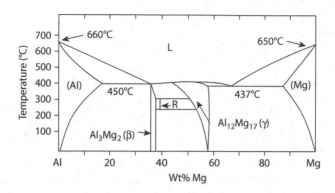

FIGURE 5.25   Al-Mg binary phase diagram [23].

layer in dissimilar metal welding confirm that the abnormality in the expansion of the interface reaction layer is due to a high rate of diffusion at elevated temperatures. IMCs like γ-Mg$_{17}$Al$_{12}$ and β-Al$_3$Mg$_2$ are persistently formed at the weld nugget zone of Al and Mg joints, as reported in many studies. These IMCs develop according to the binary phase diagram of Al-Mg combination (Figure 5.25). Thus, care should be taken during the USW process to control the temperature in order to prevent the liquation of Al and Mg metals caused by their low melting eutectic reactions.

To indicate the diffusion at the Al-Mg interface with the rise of temperature at the weld zone at every stage of weld energy, an EDS line scan analysis is the best option. Figure 5.26 reveals the appearances of bonding interfaces of AZ31B Mg/AA5052 Al ultrasonically welded joints at various levels of weld energies. It is obvious that

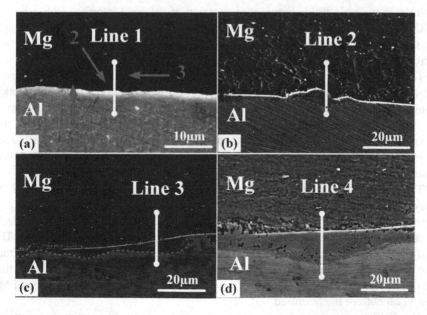

FIGURE 5.26   EDS line scan on weld cross-sections of a Mg/Al USW joint at various weld energies [17].

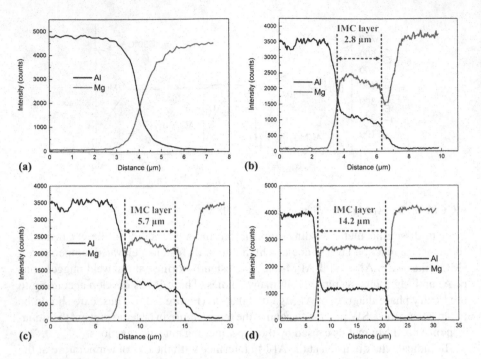

FIGURE 5.27 EDS line scan results showing diffusion across weld interface of Mg/Al ultrasonically welded joints at various weld energies [17].

no distinct IMC layer is formed at the lower weld energy (Figure 5.26(a)). However, with the increase in weld energy, some portion of the weld interface is covered by IMC islands, and the thickness of these islands is not uniform. This IMC layer gets thicker along the lateral and longitudinal directions of the weld interface with further increase in weld energy, as displayed in Figure 5.26(b)–(d).

The corresponding EDS line scan analysis of AZ31B Mg alloy-AA5052 Al alloy is presented in Figure 5.27. This analysis is useful to investigate the composition distribution along the weld interface. At the lower weld energy, the transition of Mg and Al is almost smooth, and no visible diffusion zone is noticed (Figure 5.27(a)). Nevertheless, the diffusion thickness gradually increases with the rise in weld energy values, suggesting the formation of IMCs at the Mg/Al interface (Figure 5.27(b), (c), and (d)). From the Al-Mg binary phase diagram concept, this IMC may be composed of $Mg_{17}Al_{12}$ or Mg_2Al_3.

The x-ray diffraction (XRD) analysis result of ZEK100 Mg alloy and 6022 Al alloy ultrasonic spot welded joints is displayed in Figure 5.28. This analysis was performed on fractured Al and Mg specimens to confirm the phase compositions. These fractured samples were prepared when the weld samples underwent a peel test. The analysis disclosed the existence of α-Mg+$Al_{12}Mg_{17}$ phase on both the matching fracture surfaces of the welded joint. Moreover, it can be inferred from this figure that the failure occurred through the interfacial diffusion layer due to the presence of Mg and Al on both of the fractured surfaces.

The EBSD analysis in Figure 5.29 shows microstructures and textures at the weld interface of Mg/Al ultrasonically welded joints with the variation of weld

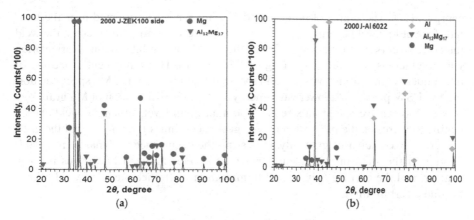

FIGURE 5.28 XRD analysis results on fractured surfaces of Al and Mg weld samples [25].

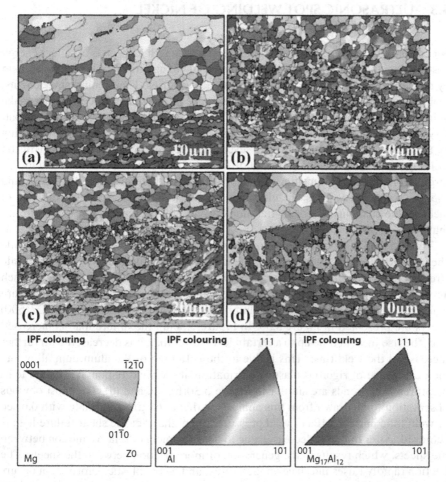

FIGURE 5.29 EBSD results at weld interface of ultrasonically welded Mg/Al joints at different weld energies [17].

energies [17]. The grains of both Al and Mg are randomly oriented near the weld interface at low weld energy (Figure 5.29(a)). However, grain refinement at the weld interface occurs when the Mg grains at the interface exhibit a definite orientation with the increase in weld energy (Figure 5.29(b)–(d)). This refinement is because of the application of normal pressure and shear stress induced in the Mg specimen during the USW process. At lower weld energy values, the distribution of Mg grain size is even and the average grain size is at the maximum level. When the weld energy begins to increase, the proportion of fine grains continues to increase and the average grain size decreases gradually, indicating the occurrence of dynamic recrystallization at the weld interface. Furthermore, the recrystallized Mg grains become coarser, and the average is increased to maximum with increments of weld energy beyond a certain limit.

5.3   ULTRASONIC SPOT WELDING OF NICKEL TO Al/Cu ALLOYS

Nickel and nickel alloys have high compressive strength as well as excellent energy absorption capabilities. Thus, a significant amount of attention has been paid to them by the aerospace, chemical, and petrochemical industries. Their remarkable benefits, including high electrical conductivity, erosion resistance, and processability, make aluminum (Al), copper (Cu), and nickel (Ni) sheets highly suitable for use in battery pole flakes, aircraft gas turbines, and packing modules. However, the joining of these Ni-Al/Cu sheets has faced several challenges in the conventional fusion welding process due to the significant differences in the metals' physical and metallurgical properties. The melting points of Al, Cu, and Ni are 660°C, 1,045°C, and 1453°C, respectively. Thus, the ultrasonic spot welding process is more efficient in tackling the challenges faced by traditional welding processes for joining dissimilar materials.

To demonstrate the relationship between the welding parameters and tensile shear failure load of ultrasonically welded joints, lap shear tensile and U-tensile strength tests are performed [26]. The effects of various welding parameters such as weld time, weld pressure, and vibration amplitude on the tensile shear failure load during the USW of AA1060 aluminum alloy-to-cupronickel (UNS C71500) sheets are presented in Figure 5.30. In Figure 5.30(a), the failure load is increased with the rise in weld time up to a certain level; after that, it is decreased with further increase in the weld time. This is due to the softening of the aluminum alloy and the occurrence of rigorous plastic deformation at the weld region at elevated temperatures. The trends are similar in Figure 5.30(b), (c), and (d). The most obvious observation to be drawn from this analysis is that at 68 µm amplitude with 0.6 sec of weld time and 0.3 MPa of weld pressure, the highest tensile shear failure load is achieved. After that, it is decreased due to the increment of relative motion between the sheets, which results in the generation of more friction between the sheets. The friction rapidly raises interface temperatures, and heavy plastic deformation occurs in the weld region.

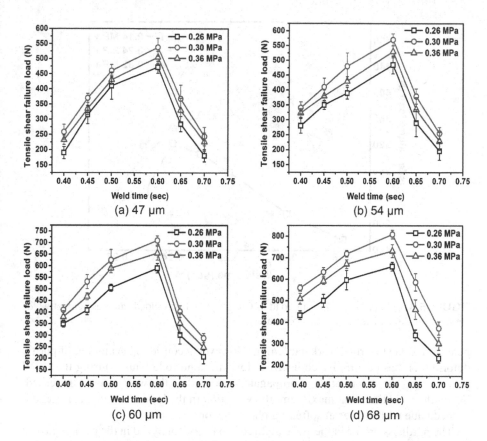

FIGURE 5.30 Tensile shear failure loads of Al-Al-(Cu-Ni) weld coupons for different vibration amplitudes [26].

Likewise, the U-tensile test is employed to investigate the mechanical performance of the ultrasonically welded joints. Figure 5.31 describes the maximum U-tensile load against the weld times for three different weld pressures of 0.3 mm pure copper (UNS C11000) and 0.7 mm nickel-plated copper samples. The U-tensile strength gradually increases with the rise in weld pressure and weld time. However, this maximum value of U-tensile strength decreases with a further increase in weld time due to the material thinning and crack formation at the periphery of the weld spot. Meanwhile, the weld samples are also not properly welded when the welding pressure and weld time are significantly less. An increase in weld time and pressure can cause the change of fracture modes from interfacial deformation to base material fracture during the USW process.

The significant amount of plastic deformation which occurs in the Cu- and Ni-coated Cu sheets is revealed by a microhardness analysis. The hardness variation horizontally along the weld spot is presented in Figure 5.32. The hardness values in the valley areas (i.e., below the sonotrode knurl) are higher as compared to the

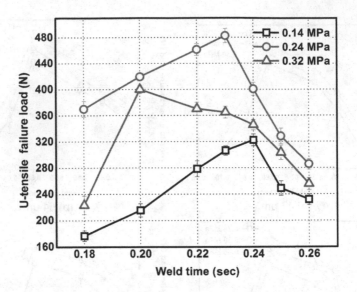

FIGURE 5.31   Maximum U-tensile load of Cu-Cu plated Ni welded samples against welding time for three different welding pressures.

peak zones due to strong work hardening. However, a considerable amount of plastic deformation has occurred specifically below the sonotrode knurls during the USW process, and the deformation is propagated outward as the weld time is increased. When the weld time is at maximum, the variation in the hardness value is lessened/reduced due to the material softening phenomenon.

The hardness profile in the vertical direction is also analyzed in the peak region to demonstrate the distribution of plastic deformed zones from the weld cross-section

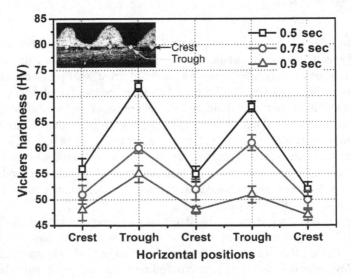

FIGURE 5.32   Horizontal hardness profile on the weld cross-sections.

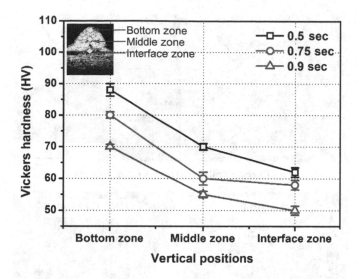

FIGURE 5.33 Vertical hardness profile on the weld cross-sections.

during the USW process (Figure 5.33). At shorter weld times, the hardness value near the surface part is much higher than at the weld interface region. This is mainly due to evident material flow at the surface region, which facilitates a huge amount of cold working in that zone. However, when the weld time is increased, the variation in hardness values no longer exists. The grains at the weld interface are elongated at shorter weld time, but at higher weld times there is no noticeable change in the grain size. Therefore, the weld samples under higher ultrasonic weld time receive more ultrasonic energy at the welding zone, and thus exhibit immediate work hardening due to cold working at the interface.

The fracture surface image in Figure 5.34 shows the indentation of sonotrode knurls, crack formation around the weld region, and various types of failure modes due to the rapid rise of interface temperature in the welded specimens. The failure modes are mainly of three types depending on the weld time. At low weld time, the imprint of the sonotrode tip is shallow, and less weld energy is transmitted to the weld region. However, the imprint of the sonotrode tip on the weld specimen is deeper when the weld time is increased. As a result, the weld energy of the weld interface is also increased. This results in the changing of direct failure mode to transverse through failure mode and nugget pull-out failure mode at different weld times. The maximum tensile strength of joints can be achieved in transverse through failure mode because continuous microbond formation occurs at the weld interface. The nugget pull-out failure mode is different from the other two modes due to the thinness and softening of the top metal sheet. As a result, a crack has occurred at the weld spot at higher weld time.

Optical microscopy images of different weld samples (such as under, good, and over weld specimens) can be analyzed to demonstrate various weld qualities. Cross-sectional images of Cu and Ni-plated Cu at different weld times are shown

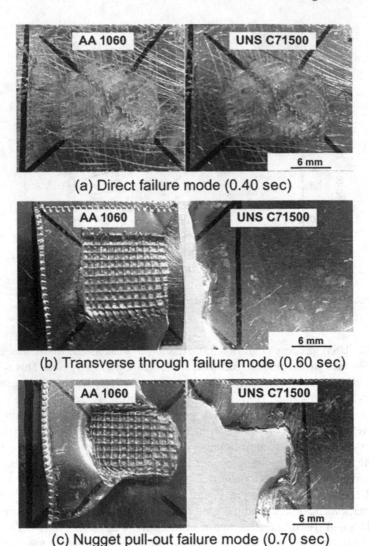

(a) Direct failure mode (0.40 sec)

(b) Transverse through failure mode (0.60 sec)

(c) Nugget pull-out failure mode (0.70 sec)

FIGURE 5.34   Various failure modes of weld specimens at different weld times [26].

in Figure 5.35. The gap is clearly visible along the bonding line at low weld time. Nevertheless, when the weld time increases, these gaps become smaller and are fully covered by the microbonds. At higher weld time, several cracks are observed in the weld interface due to the diffusion across the weld zone, local melting, and mechanical interlocking at the weld interface. The shape of the local microbonds is changed in the cracked Ni layer due to continuous shear oscillations around these bonds. This bonding line has a twisted and rolled shape; thus, it is called an interfacial wave. Nevertheless, the densities of metallic bonds and interlocking in the over weld are much higher than in the good welds. Therefore, the weld performance of joints is not a function of bond density alone.

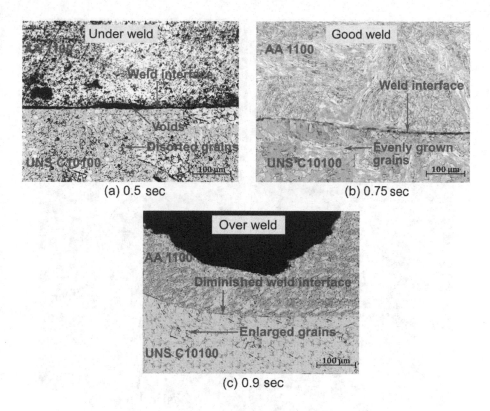

FIGURE 5.35 Optical images of ultrasonically spot welded joints showing three different weld qualities.

The electron microscopy images of the Al-Ni foam joint fracture surface in Figure 5.36 reveal two sections that have undergone different types of fractures. In this figure, Al and Ni foam sheets are presented as A and B sections respectively. Section A exhibits the bonding region between the Al sheet and fractured parts of the Ni foam sheet under the high shear stress generated in this weld area. Meanwhile, section B shows a three-dimensional (3D) triangular strut type of fracture under the tensile stress generated in this region. The indentations of the sonotrode knurls are rhombic in nature, as is clearly illustrated in Figure 5.36(b). It is evident from this figure that most of the 3D triangular struts are able to retain their complete structure and consistency under the plastic deformation. Apart from the plastic deformation zone, there are some zones where only grinding scratches are observed. These scratches are parallel to the vertical vibration direction, and they are quite distinct on the aluminum side (Figure 5.36(c) and (d)). Figure 5.36(e) and (f) displays the hollow struts which are bounded by nickel-grain walls. Some plastic slip lines, necking areas, and dimple-free fractures are also observed along with the strut walls.

Furthermore, a field emission scanning electron microscopy (FESEM) analysis along the bond-line of welded samples at different weld times is shown in

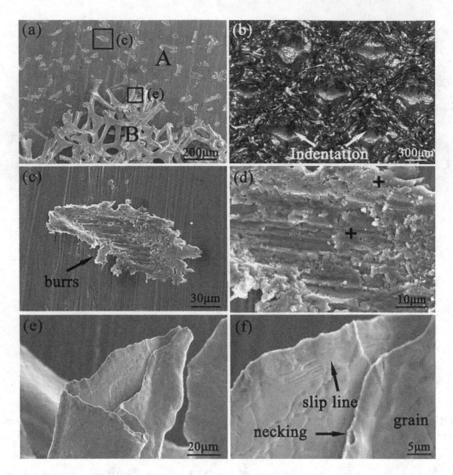

FIGURE 5.36   (a) SEM images of fractured surfaces of ultrasonic spot welded Al sample, (b) indentations on the deformed nickel foam sheet, (c) magnified view of the region shown in (a), (d) central zone of (c), (e) magnified image of fractured struts shown in (a), and (f) central zone of (e) [27].

Figure 5.37. The dotted parts in the figures are magnified and presented in Figure 5.37(b), (d), and (f). At the weld region of the under weld sample, the gap is clearly visible. However, when the weld time increases, the gap becomes smaller and is fully covered by microbonds. Various bonding zones with similar grain structure are discovered at the weld spot in the good and over weld samples. The welded samples are in perfectly intimate contact with each other at good weld conditions. As compared to good weld conditions, the density of the metallic bond is abundant in over weld samples at higher weld time. Figure 5.37(f) shows the presence of several interlocking features at the bonding interface, which provide some extra weld strength to the joints. Thus, this bond density and mechanical interlocking formation are the major sources of the good weld performance in USW of Al-CuNi samples.

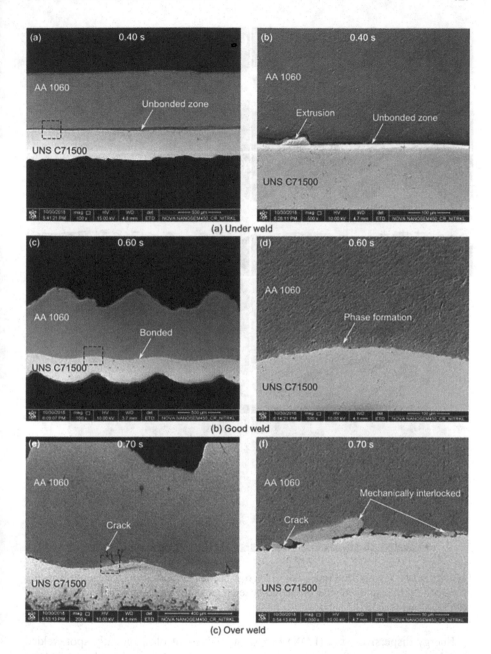

(a) Under weld

(b) Good weld

(c) Over weld

FIGURE 5.37 FESEM images of ultrasonically spot welded Al-CuNi joints showing different weld qualities [26].

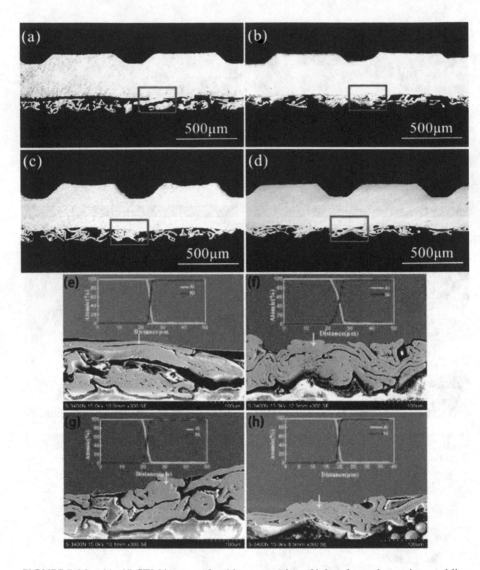

FIGURE 5.38 (a)–(d) SEM images of weld cross-section of joints formed at various welding energies; (e)–(h) magnified view of the red-boxed regions shown in (a)–(d) respectively done with EDX line scan [27].

Energy dispersive x-ray (EDX) line scan analysis of ultrasonically spot welded Al-Ni joints can reveal the diffusion-related mechanisms. Figure 5.38 shows the joint interface between Al and Ni form sheets, which reveals insufficient plastic deformation and material flow at low weld energy. When the weld energy is increased, the 3D triangular struts are compressed and the inter-strut distance decreases. Thus, the contact area between the weld interfaces is improved. The maximum weld energy results in the formation of a noticeable contact area between the two welded sheets, where the nickel foam has become excessively thin. Some interdiffusion area is also

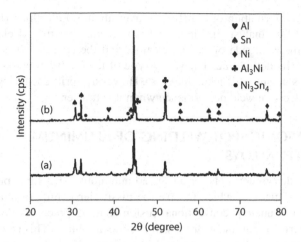

FIGURE 5.39 XRD patterns of ultrasonically spot welded Al alloy joints with (a) Ni80-Sn and (b) Ni60-Sn foils [28].

clearly visible across the weld interface. Smooth composition transition indicates that only low amounts of solid solution and IMCs have formed at the weld spot.

Figure 5.39 explains the XRD patterns of ultrasonically welded Al alloy with Ni80-Sn and Ni60-Sn foil joints. Peaks of Ni_3Sn_4 phase and Al_3Ni phase, along with those of Al, Ni, and Sn elements, are indexed in the patterns. These results convincingly demonstrate that Ni_3Sn_4 and Al_3Ni IMCs are formed in both cases.

Enhancement in the mechanical performance of the Al-Ni joints can also be achieved by the addition of an interlayer in the form of sheet or powder. When Al2219 powder is added as an interlayer between the Al and Ni sheets, the friction increases, leading to an increase in temperature at the weld interface (Figure 5.40(a)). Therefore, the weld specimens become softened, and the yield strength decreases significantly. However, the high temperature is favorable for effective material flow at the weld zone, and more fresh metal-to-metal contact area has formed. In contrast, inadequately combined interfaces occur at the early stages of the USW process. The TEM analysis of the interface of the Al2219 powder interlayer and pure Ni sheets

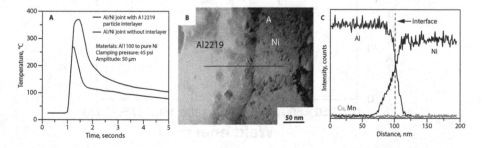

FIGURE 5.40 (a) Interface temperature variation with weld time; (b) TEM image; (c) EDX analysis results of ultrasonically spot welded Al2219 powder interlayer and pure Ni sheets [29].

demonstrates the formation of a diffusion layer about 60 nm thick (Figure 5.40(b) and (c)). The EDX analysis in Figure 5.40(c) reveals that the Al element sharply decreases on the right side of the dotted line, and the Ni element is decreased on the left side of the dotted line. The valley type of feature between these two dotted lines represents the interdiffusion thickness. However, a brittle IMC layer always has detrimental effects on weld performance when it is thicker.

5.4 ULTRASONIC SPOT WELDING OF ALUMINUM TO STEEL ALLOYS

Multi-material design concepts for advanced automotive structures provide weight-saving solutions, many of which involve the dissimilar joining of aluminum alloys with steels. Unfortunately, conventional fusion welding processes do not presently meet the industrial requirements for dissimilar metal joining. This is mainly because a high rate of intermetallic reaction occurs in the liquid phase, leading to poor properties and low joint strength. USW is a comparatively new and efficient solid-state process that could potentially be substituted for resistance spot welding, and it can reduce the tendency toward intermetallic reaction because there is no liquid phase formation.

Tensile lap shear strength tests are employed to analyze the mechanical strength of the ultrasonically spot welded Al6061-T6 Al alloy to galvanized HSLA steel and Al6061-T6 Al alloy to ASTM A36 steel. This strength is calculated as the tensile lap shear load divided by the sonotrode tip area. The maximum lap shear strengths of these dissimilar joints are plotted against the welding energy at constant power and clamping pressure (Figure 5.41). The tensile lap shear strength of Al alloy to ASTM

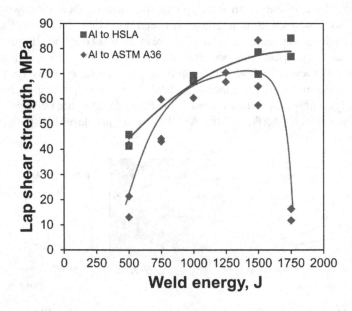

FIGURE 5.41 Variation of lap shear strengths of ultrasonically welded Al 6061-HSLA steel joints and Al 6061-ASTM A36 steel joints at various welding energy values [30].

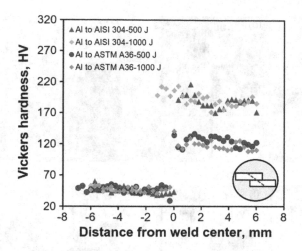

FIGURE 5.42 Distribution microhardness values at the weld interface of dissimilar ultrasonically welded Al 6061-AISI 304 steel and Al 6061-ASTM A36 steel joints at various welding energies [31].

A36 steel dissimilar joints first increases with the increase in energy inputs and reaches a peak value at 1500 J weld energy. Thereafter, it decreases with a further increase in the welding energy. At lower weld energy, the temperature is not high enough to soften the Al-steel interface. In contrast, at higher weld energy the weld sample is subjected to higher temperatures under larger vibration amplitudes for a longer time. As a result, a brittle intermetallic layer (composed of $FeAl_3$ and Fe_3Al phases) may be present at the weld interface, and it decreases the weld strength of these dissimilar joints. However, this trend is different for Al to galvanized HSLA steel joints, where the tensile lap shear strength increases with the increasing energy inputs and achieves the highest value at 1750 J.

Figure 5.42 shows the microhardness at the weld interfaces of dissimilar ultrasonically spot welded Al to AISI 304 stainless steel and Al to ASTM A36 steel joints at various levels of welding energy. These were made to study the effect of natural aging and softening across the weld region. One can see that no noticeable HAZ is present for either joint. Noticeable asymmetrical-type hardness profile across the weld zone for both dissimilar joints is obtained with an average hardness value on the Al and steel side. The figure shows that AISI 304 stainless steel possesses a higher hardness value than ASTM A36 steel overall. This is likely due to the high strength of the bulk AISI 304 stainless steel and the solid solution strengthening of Cr. Generally, the hardness value starts to decrease with a rise in welding energy due to the generation of high interface temperature, which ultimately leads to an increase in grain size. As shown in Figure 5.42, the hardness values in the Fe side of ASTM A36 show the decreasing phenomenon, whereas it is not distinguishable in the AISI 304 Fe side.

The specific reason for the tensile lap shear strength variations can be perceived from the fracture surface morphology. Figure 5.43 presents the fracture surfaces of tensile lap shear tested samples at different energy inputs. An evolution of failure

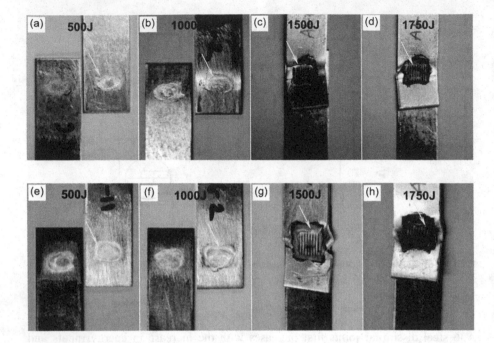

FIGURE 5.43 Morphology of fracture locations for ultrasonically welded ((a)–(d)) Al 6061-HSLA steel and ((e)–(h)) Al 6061-ASTM A36 steel [30].

mode around the weld region can clearly be detected when the welding energy is increased. A tensile lap shear fracture occurs in the Al-Fe region when lower weld energy is applied to both Al to galvanized HSLA steel (Figure 5.43(a) and (b)) and Al to ASTM A36 steel joints (Figure 5.43(e) and (f)). The interfacial deformation with the micro-welds at the welded interface is clearly visible on both Al and Fe sides at low energy input. When the welding energy is increased, the size and number of micro-welds are also increased. In spite of this, the interface temperature at the weld interface is not high enough to severely deform the steel specimen or diffuse the Zn interlayer into the Al to achieve an over weld joint. Hence, a weld pull-out failure mode occurs at this weld energy, where the bonded region at the weld interface requires a large amount of load to be separated, as shown in Figure 5.43(c) and (g). Meantime, at higher weld energy, the weld samples for both joints are subjected to higher temperatures under larger vibration amplitudes for a longer period of time. This leads to an increase in interatomic diffusion between Al and steel joints, and the Zn interlayer is squeezed out for galvanized Al-HSLA steel joints. With such high-energy inputs, the failure generally occurs at the edge of the nugget zone on the softer Al side, and it is referred to as transverse through-thickness crack growth (Figure 5.43(d) and (h)). To summarize, the failure mode changes from interfacial shear failure to transverse through-thickness failure in the base metal. This twisting phenomenon leads to a small micro-level crack at the notch of two welded sheets, resulting in a higher stress concentration in that zone. As a result of stress, the remaining cross-section can no longer sustain the shear load. The heavily deformed

Al sheet eventually allows cracks to grow in the transverse direction, and thus it experiences through-thickness crack growth. It should be noted that welded joints made at a higher welding energy inputs may be stronger, but the overall strength of the sample is limited due to the higher stress concentration at a thinner nugget edge.

Optical microscopy on the weld cross-sections reveals the quality of welds produced during the USW process. Figures 5.44 and 5.45 show optical micrographs of dissimilar ultrasonically welded Al to AISI 304 stainless steel and Al to ASTM A36 steel joints at various welding energy values. At lower weld energy, it can be seen

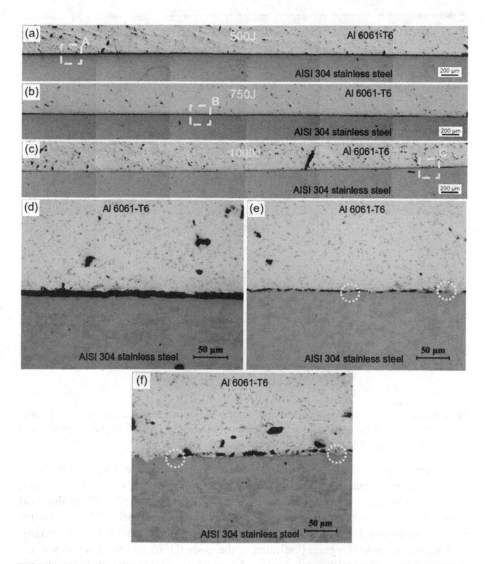

FIGURE 5.44 (a)–(c) Overall optical microscopic views of ultrasonically welded Al 6061-AISI 304 steel joint cross-section at different weld energies; (d)–(f) higher-magnification images of the dotted boxes in (a)–(c) respectively [31].

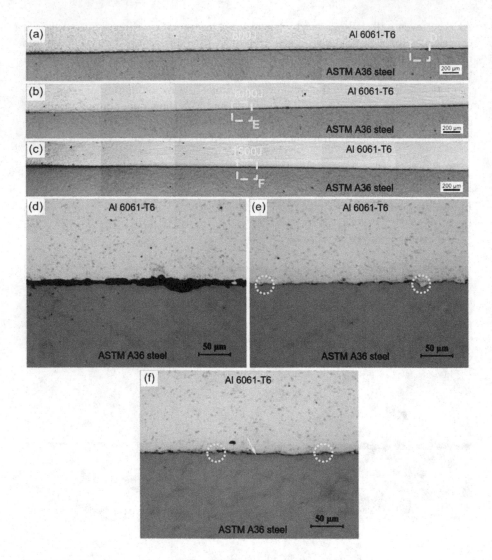

FIGURE 5.45 (a)–(c) Overall optical microscopic views of ultrasonically welded Al 6061-ASTM A36 steel joint cross-section at different weld energies; (d)–(f) higher-magnification images of the dotted boxes in (a)–(c) respectively [31].

from Figures 5.44(a) and 5.45(a) that gaps exist along the weld region, demonstrating a lack of bonding between the two samples. When the welding energy is increased, the penetration of the deformation is also increased significantly. When the weld energy is increased, the gaps at the weld region become less distinct (Figure 5.44(b) and (c) and Figure 5.45(b) and (c)). This is made more understandable by the magnified images of various regions (indicated by the dashed boxes) of welded joints at different weld energies, as shown in Figures 5.44(d), (e), and (f) and 5.45(d), (e), and (f). At the higher weld energy, no void is present in the weld region (Figure 5.44(f) and 5.45(f)). As a result, a clear continuous line can be seen in the weld region.

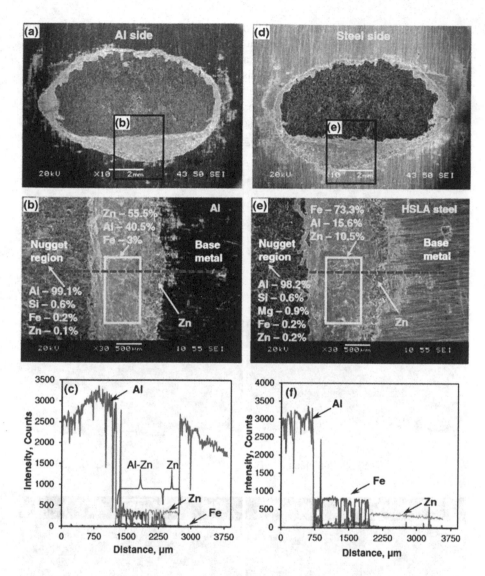

FIGURE 5.46 Fractured-surface SEM photographs of ultrasonically welded Al-HSLA steel joints: (a) overview of Al surface, (b) magnified view of the boxed section of (a), (c) EDS line scan analysis results of Al side, (d) overview of HSLA steel surface, (e) magnified view of the boxed section of (b), (f) EDS line scan analysis results of HSLA steel side [32].

Figure 5.46(a) and (d) show the overall SEM images of a fracture surface on the Al side and steel side. SEM analysis was performed on the welded tensile samples, which failed in an interfacial mode at maximum welding energy. There are two different sections, as the weld center and weld edge are observed on both the Al and the steel sides. Figure 5.46(b) is a magnified view of the box in Figure 5.46(a) on the Al side, exhibiting the weld nugget, weld edge, and base metal. Likewise, Figure 5.46(d) shows the Al-Zn liquid interlayer on the steel side in which the melted Zn has been squeezed out from the weld nugget and formed an Al-Zn eutectoid and eutectic layer. The squeezing of the melted Zn layer is attributable to the lower melting point of Zn as compared to Al. Additionally, Zn has a higher vacancy concentration during the USW process. An EDS line scan analysis was also performed on the Al and steel sides, and the results are shown in Figure 5.46(e) and (f).

To identify the weld attributes more meticulously at different levels of weld energy, higher-magnification SEM back-scattered electron images of the weld interface of dissimilar joints are required (Figure 5.47). There are no huge defects like

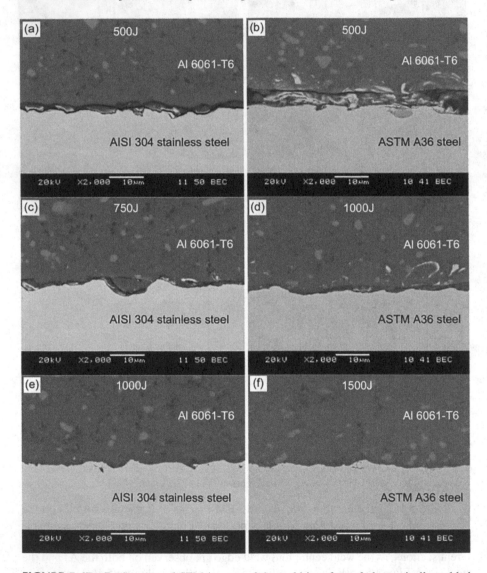

FIGURE 5.47 Back-scattered SEM images of the weld interface of ultrasonically welded ((a), (c), and (e)) Al 6061-AISI 304 steel alloy and ((b), (d), and (f)) Al 6061-ASTM A36 steel alloy at different weld energies [31].

gaps, cracks, or tunnels present in the welded samples under most of the welding conditions except for the lower-energy joints. When the weld energy increases, there is a considerable increase in the interface temperature, which softens the weld specimen. Thus, the sonotrode knurls sink deeper into the sheet, and the bond line is displaced to a wavelike slow pattern, as depicted in Figure 5.47(c)–(f).

Figure 5.48 displays EDS line scans at the weld interface of Al-HSLA steel and Al-ASTM A36 steel joints to show the various phases formed during the USW process. It can be seen that the interface condition for both types of joints is not similar from a microstructural point of view. In the case of Al-HSLA steel, the concentration of Al starts to decrease around 4 μm, as shown in Figure 5.48(c). However, Figure 5.48(a) reveals that the interface of the Al-HSLA steel joint possesses a larger concentration of Al and Zn than of Fe. According to the binary phase diagram of Al-Zn and EDS line scan analysis, region 1 marked in Figure 5.48(a) may be composed of Al-Zn eutectic film. The interface temperature for Al-HSLA steel is higher than the eutectic melting point of Al-Zn. Thus, a thin film of Al-Zn eutectic liquid is expelled from the center of the weld zone to the weld periphery. In contrast, the interface between Al-ASTM A36 steel is fairly straight with no eutectic film present (Figure 5.48(b)). In this joint, the Al concentration begins to decrease at around 4.5 μm, and the Fe is

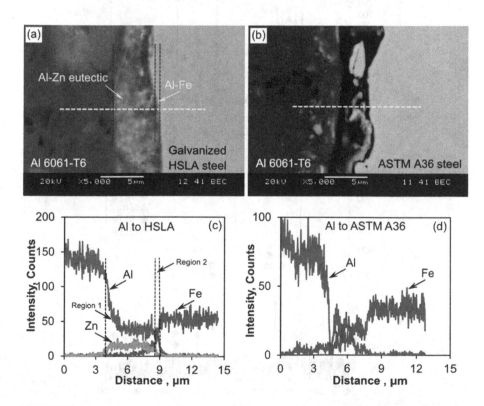

FIGURE 5.48 EDS line scan analysis with SEM images of the weld interface showing diffusion in ultrasonically welded (a) Al 6061-HSLA steel and (b) Al6061-ASTM A36 steel joints [30].

merged more in the Al than in the Al-HSLA steel joint (Figure 5.48(d)). Furthermore, it is observed that the interdiffusion layer thickness is higher in Al-ASTM A36 steel than in the Al-HSLA steel joint. As the binary phase diagram of Al and Fe shows, Al has a greater solubility in Fe up to ~11 wt% of alloying. However, if the concentration of Al is higher than this critical limit, then brittle IMCs like $FeAl_3$, Fe_2Al_5, $FeAl_2$, and FeAl may evolve, which ultimately deteriorates the strength of the weld.

It is evident from the XRD analysis of the fractured surfaces of Al and steel that the highest peaks of Al, Fe, Zn, Fe_2Al_5, and $FeAl_3$ phases are formed during the USW process (Figure 5.49). These brittle IMCs are formed when the pure metal-to-metal interaction happens, and the phase which has the most negative Gibbs energy will be

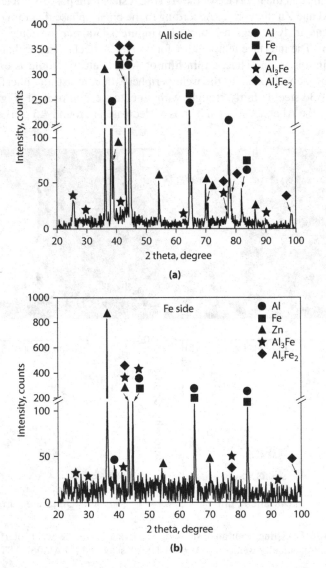

FIGURE 5.49 XRD analysis results on fractured surfaces of (a) Al side, (b) steel side [33].

kinetically favored. Based on this phenomenon, the FeAl$_3$ phase has the least effective free energy, and thus it is kinetically favored. After the formation of this phase, it may react with Fe to form an Fe$_2$Al$_5$ phase.

Figure 5.50 displays the EBSD mapping of the phases formed with the IMC interlayer during USW between Al-steel samples. In Figure 5.50(a), the columnar

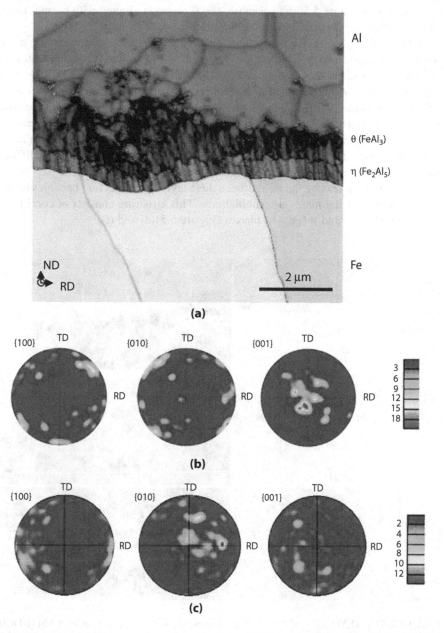

FIGURE 5.50 EBSD mapping of dual-phase IMC reaction layer in AA6111-DC04 steel sample [34].

grain structures of IMC phases and tiny particles of the $FeAl_3$ phase appear in the Al matrix. The pole figures are presented in Figure 5.50(b) and (c), where RD denotes the rolling direction of the sheet, and TD represents the transverse direction. The highest Al content phase (θ-$FeAl_3$) is readily formed in the Al side, whereas the highest Fe content phase (η-Fe_2Al_5) is formed in the Fe side. Furthermore, these pole figures reveal the η phase direction oriented normal to the interface plane direction, as seen in Figure 5.50(b). Likewise, the θ phase is aligned with normal to the interface plane, as portrayed in Figure 5.50(c).

A TEM micrograph is the best option to develop a complete understanding of the bonding mechanism in the IMC region. Figure 5.51(b), (d), and (f) demonstrates the IMC reaction layer at the interface of AA6111-DC04 steel and AA7055-DC04 steel joints at different weld times. The dotted lines split up the different phases. When the welding time is less, the IMC zone is wide, and a single grain is observed at the bond line of the interface. When the welding time increases, the IMC layer is converted to an irregularly thick, continuous layer. Figure 5.51(b) depicts the IMC growth in the interface region, and it is identified through TEM by use of the selected area diffraction (SADP) technique. This interface is completely composed of η (Fe_2Al_5) phase. However, upon increasing the weld time, a dual-layer IMC structure becomes recognizable in both of the material combinations. This structure consists of continuous layers of θ ($FeAl_3$) and η (Fe_2Al_5) phases (Figure 5.51(d) and (f)).

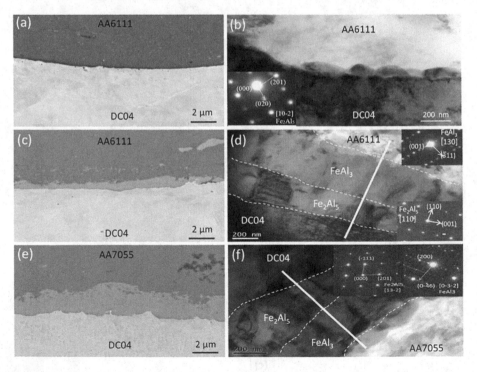

FIGURE 5.51   TEM images of the IMC layers along the weld cross-section for AA6111-DC04 steel and AA7055-DC04 steel joints at various weld times [34].

5.5   ULTRASONIC SPOT WELDING OF ALUMINUM TO TITANIUM ALLOYS

Improvement in the fuel economy and reduction in the gross weight of an automobile can be obtained through multi-material parts produced from titanium and aluminum alloys. USW meets the industrial requirements for this type of dissimilar metal welding due to its low welding temperature and little energy input, which avoids the formation of a thick and brittle intermetallic layer at the weld region. Therefore, the USW process is employed to produce most of the successful Al-Ti joints, because of its solid-state nature.

The mechanical strength of ultrasonically welded samples can be evaluated through the tensile shear test. Figure 5.52 demonstrates the relationship of shearing failure force to various welding times and welding pressures of the titanium (Ti6Al4V) and Al (AA6061) samples. When the shearing force is applied to the weld samples, the failure tends to proliferate around the weld interface. This implies that the weld samples attain a strong bond at the nugget zone, and shows the strength of the joint over the strength of the aluminum alloy. The shearing failure force increases gradually with the increase in welding time under different static pressures. The shearing failure force increases up to a certain limit, and after that it decreases. Meanwhile, at the initial state, the weld strength is lessened by a short welding time or the presence of residual oxide at the weld region. At higher weld time, these samples are subjected to higher temperatures under larger vibration amplitudes for a longer time. On the one hand, when the temperature of the sample is increased, softening of the aluminum alloy occurs, and more plastic deformation occurs at the weld interface. However, when the welding time is too long, crack formation takes place at the periphery of the weld surface. The ideal welding time is always dependent on the properties, thickness, and other process parameters of the weld material.

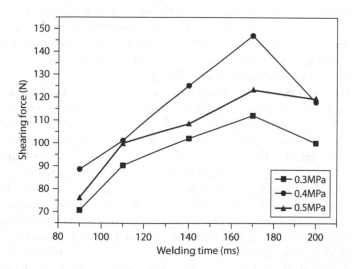

FIGURE 5.52   Variation of shearing failure force with weld time and weld pressure for ultrasonically welded Al-Ti joints [35].

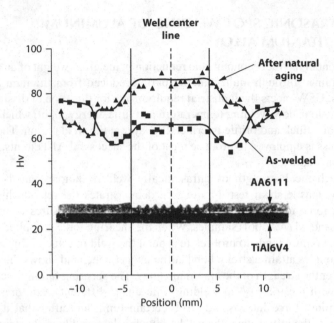

FIGURE 5.53 Comparison of microhardness profiles across the weld cross-section of AA6111-Ti6Al4V dissimilar alloys just after the USW process and after four days [36].

Figure 5.53 compares the microhardness of the weld cross-section during the USW of AA6111 Al alloy and Ti6Al4V with the same cross-section after four days of natural aging. The variation trend of these two hardness profiles exhibits similar characteristics under different welding times. The maximum hardness in the weld interface after natural aging exceeds the original microhardness of the weld cross-section, which is ~80 HV. Moreover, the minimum hardness value is observed in the thermo-mechanical affected zone (TMAZ) of the aluminum alloy surface after welding, and it is eliminated by the natural aging process. After four days of natural aging, the hardness is recovered partly due to re-precipitation of the alloying elements that were dissolved during the welding process.

The influence of penetration of sonotrode knurls, the formation of cracks around the weld spot, and the quick rise in interface temperature lead to various types of fracture surfaces in welded samples. The fracture surfaces of the welded samples at different welding time are shown in Figure 5.54. When the welding time is at minimum, the imprint of the sonotrode tip is very small, due to the transfer of the low amount of weld energy to the weld interface. Meantime, different sizes of micro-welds are formed by sliding friction at the faying surfaces of Al-Ti alloy sheets. However, the imprint of the sonotrode tip on the weld samples increases with an increase in weld time. For welding time longer than 1 sec, the bond area spreads over the whole joint. In all cases, the failure has occurred at the weld interface, and this type of failure is called a *pull-out failure mode*. This is because of the nonsubstantial plastic deformation of the AA6061 aluminum alloy during the welding process.

Optical microscopy (OM) on the weld cross-section reveals the quality of weld produced during the USW process. OM images of the ultrasonically welded Al-Ti samples are shown in Figure 5.55. The top face of the aluminum sheet is softened

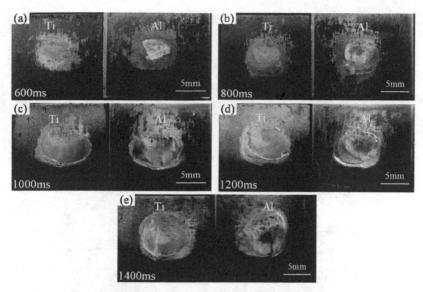

FIGURE 5.54 Morphologies of fracture surfaces of Al-Ti dissimilar joints at various weld times [37].

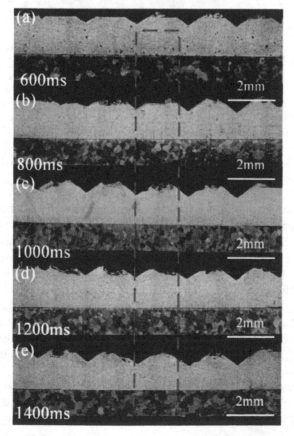

FIGURE 5.55 Optical microscopy of Al-Ti weld cross-section made at different welding times [37].

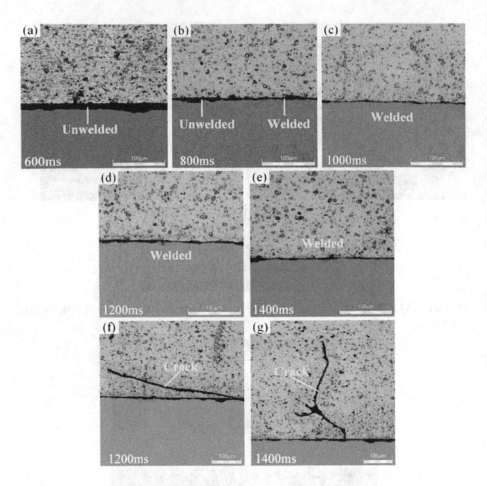

FIGURE 5.56 Magnified optical microscopic view of weld interface of Al-Ti ultrasonically welded dissimilar joints at various weld times [37].

by the heat generation due to the shearing action between the sonotrode and the top sheet, and it tends to stick on the sonotrode tip as the welding time is increased. This is more understandable from the magnified images of various regions (indicated by the dashed boxes) of the weld joints at different weld times, as shown in Figure 5.56(a)–(g). When the welding time is too short, gaps are clearly visible along the weld interface, which shows the lack of bonding between the two samples. Whenever the welding time is increased, the deformation due to the penetration is also increased significantly, and it results in lessening the gaps at the weld interface. At the highest weld time, the ultrasonic vibration duration is also too long, and due to that, fatigue crack occurs on the aluminum surface of the joint, as shown in Figure 5.56(f) and (g).

A SEM investigation with EDX line scan analysis for fracture surface conditions is shown in Figure 5.57. The fracture surface of the Al-Ti samples demonstrates no considerable difference in the morphology at the weld center as compared to the

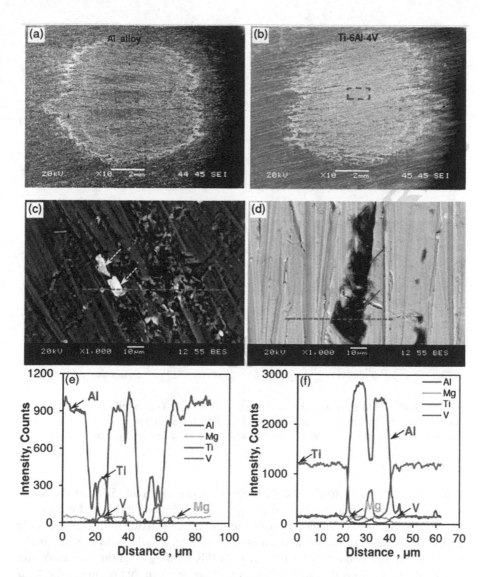

FIGURE 5.57 SEM images of the (a) Al fractured surface, (b) Ti fractured surface; (c) and (d) magnified views of the dotted boxes in (a) and (b) respectively; (e) and (f) EDX line scan analysis across weld interface as indicated by the dashed lines in (c) and (d) [38].

edge of the nugget region at low magnification (Figure 5.57(a) and (b)). Some small white fragments (indicated by the dashed arrow in Figure 5.57(c)) on the Al surface are noticeable on the magnified view of the fracture surface, and these are rich in Ti and V elements as confirmed by the EDS line scan (Figure 5.57(e)). Likewise, a few grey discontinuous "islands" (indicated by solid arrows in Figure 5.57(d)) on the Ti surface are detected, which are mostly the Al element as confirmed by the EDS line scan (Figure 5.57(f)). Results from another study using SEM analysis of

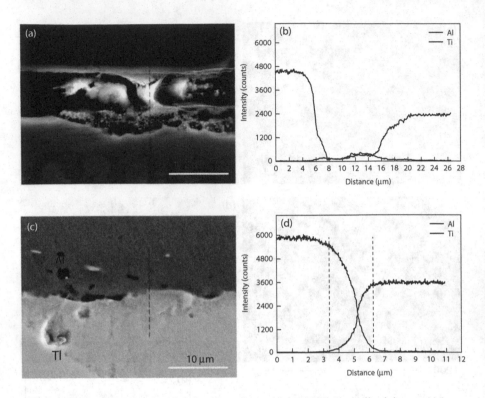

FIGURE 5.58 SEM images of the weld cross-section of Al-Ti dissimilar joints at (a) lower welding time, (c) higher weld time; (b) and (d) EDS line scan analysis showing diffusion across the weld interface as indicated by the dotted line in (a) and (c) respectively [37].

weld cross-sections at different weld times is shown in Figure 5.58. At the lower welding time, there is a clearly visible gap (Figure 5.58(a) and (b)), and it is due to the fact that the weld time was not sufficient to break the oxide film. Figure 5.58(c) depicts the weld cross-section at the higher weld time, and the weld samples are in perfectly intimate contact with each other. At this welding condition, the EDS line scan analysis confirms the atomic diffusion of about 3 μm that occurs at the weld interface (shown in Figure 5.58(d)). This result confirms that no IMC or significant interfacial reaction layer is formed in the weld zone, and it can be clarified from the solid-phase bonding characteristics of Al-Ti dissimilar joints during ultrasonic welding. Meanwhile, IMCs can hardly form along the Al-Ti weld cross-section at the lower weld energy, as this energy is not sufficient to activate diffusion between the sheets. However, longer welding time between Al and Ti leads to the formation of hard and brittle IMCs, as shown in the Al-Ti binary phase diagram (Figure 5.59) of the weld region.

XRD analysis can be performed to classify the phase of IMC formation in the joints during the USW process. Figure 5.60(a) and (b) confirms that the XRD pattern of the Al and Ti phases appeared on the Al and Ti sides, respectively. No IMCs are

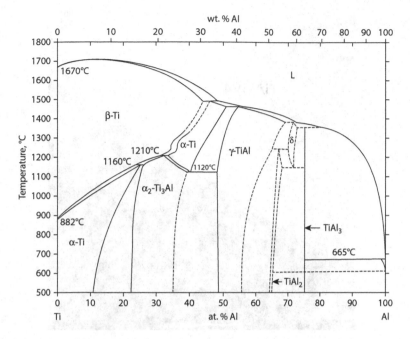

FIGURE 5.59   Al-Ti binary phase diagram [39].

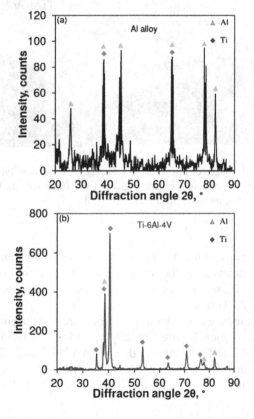

FIGURE 5.60   XRD analysis results of dissimilar ultrasonically welded Al-Ti joint [38].

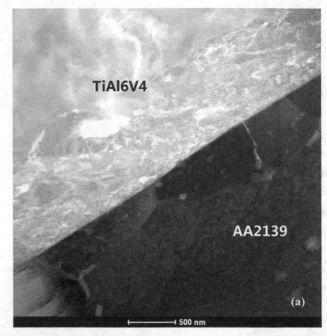

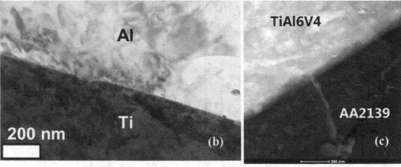

FIGURE 5.61 TEM images of ultrasonically welded Al-Ti dissimilar weld interface: (a) a low-magnification image showing a large region of the interface, (b) higher-magnification bright-field image, (c) higher-magnification image [40].

detected in the interface of the Al-Ti joints. USW can be an appropriate method for Al-Ti dissimilar joining due to the reduced tendency to form brittle IMCs, such as Ti_3Al, TiAl, and $TiAl_3$, which are commonly formed during dissimilar fusion welding and brazing of Al-Ti sheets.

TEM analysis can be used at the weld region to reveal the complete microstructures at a higher resolution. Figure 5.61 displays bright-field and annular dark-field images at the weld interface between the Al (AA2139) and Ti (TiAl6V4) specimens. At the higher resolution of TEM images, no IMC layer is noticed. If any layer is present, it must be very skinny (less than 100 nm). Thus, it is considered negligible.

5.6 ULTRASONIC SPOT WELDING OF ALUMINUM TO BRASS ALLOYS

The major significance of a good joint between dissimilar materials like aluminium and brass find its numerous applications in the electrical industries. Generally, copper and brass are utilised extensively in these industries because of its superb electrical and thermal properties, high strength and erosion and fatigue resistances. But, as the copper has higher thermal conductivity, thermal expansion and electrical conductivity than brass thus, a larger weld distortion occurs in analogous to brass welds. Thus, it is more important to examine the weld characteristics of joining of aluminium to brass. In friction stir welding technique, whenever brass was tried to weld with aluminium, hard and brittle intermetallic compounds (IMC) were formed giving poor weld strength. Meantime, it is important to pursue lower cost joining methods. Therefore, USW is believed to be one of such promising method for joining these type of softer metals to overcome the associated difficulty. The weldability of brass mostly relies upon the percentage of zinc present in brass. As the zinc has a low boiling temperature, lethal vapours may produce from these kinds of welding processes.

It has already been discussed that the ultimate tensile shear and T-peel failure loads are the functions of weld pressure, weld time and vibration amplitude. Figure 5.62

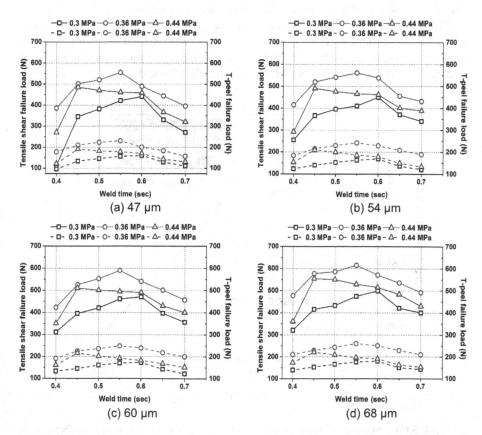

FIGURE 5.62   Tensile shear and T-peel failure loads of Al-brass weld coupons for different vibration amplitudes.

represents the tensile shear strength and T-peel failure loads for Al-Brass specimens at various amplitudes. In this figure, the solid lines represent tensile shear failure loads, and dotted lines show T-peel failure loads. In these experimental results, one common feature is noticed, i.e., the failure loads of various welds first increase and afterwards diminish with an increase in weld time. It is believed that the cracks around the weld spot may be formed due to prolonged weld time. Meanwhile, at the moderate amount of weld pressure and at the highest amplitude of vibration, the scrubbing action between the faying surfaces increases. Thus, it results in a better bonding strength. From this observation, it is clear that each control parameter has a separate and significant influence on the weld strengths. Moreover, it is noticed that when the thickness of the weld coupons increases, the strength of the joint also increases.

Thus, the microhardness measurements on the cross-sections of "under," "good," and "over" weld samples reveal the work hardening and softening during the welding process. Vickers hardness results plotted against the weld time are shown in Figure 5.63. From this figure, it is observed that at the starting stage of USW, the hardness shows 48% higher results than the hardness of the parent materials, due to severe cold working at the interface. When the weld time is increased further, the hardness value is drastically reduced due to material softening.

The hardness varies along the weld interface in a horizontal manner with respect to various weld times for these three weld conditions, as shown in Figure 5.64. For the lowest weld time, the microhardness under the peak of the sonotrode is higher than at the crest position. It is due to the fact that the plastic deformation starts from

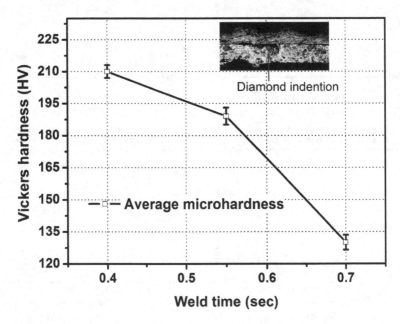

FIGURE 5.63 Average microhardness of peaks beneath sonotrode tip for different weld times.

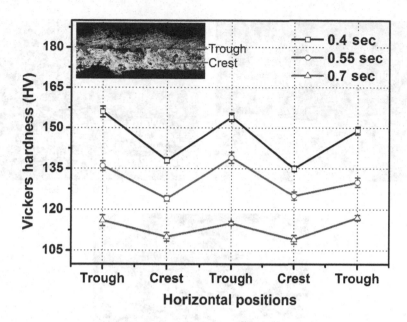

FIGURE 5.64 Hardness distribution along the horizontal direction for Al-brass weld samples.

the center of the peak position and gradually disperses toward the crest position as the welding proceeds. Thus, the crest region is not as work hardened as is the material in the peak position. Furthermore, for a shorter weld time, the fluctuation in hardness value is more than that seen with higher weld time.The hardness value is also lowered at the highest weld time.

In USW, the microstructural changes occur initially due to the plastic deformation and work hardening of the weld specimen, followed by softening due to the generation of high temperature with the increase of weld time. Due to the work hardening, the initial grain size of the parent material elongates, so grains are noticed only in the initial stage. As the time increases further, when recrystallization temperature is reached new fine grains are formed, and the formerly elongated grains disappear. The newly formed grains have the same lattice structure as the undeformed grains of parent material. However, if the welding is continued further, then the grains begin to grow, and consequently the hardness in this region decreases. Microscopic pictures of grains around the weld interface with respect to various weld times are shown in Figure 5.65.

For the microstructural analysis of ultrasonically welded Al-brass, three types of test specimens can be selected based on the tensile shear and T-peel failure loads occurring at the highest vibration amplitude. Figure 5.66 shows solid-state diffusion between the weld surfaces in SEM images along the cross-sections of "under," "good," and "over" weld conditions. In the under weld condition, a void between the two weld surfaces is spotted, and no microbonds are found. In the good weld condition, the weld interface is tightly packed with microbonds and

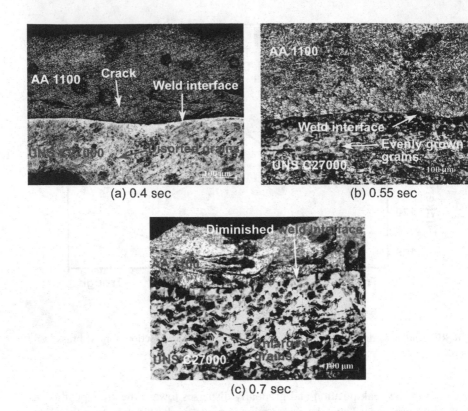

(a) 0.4 sec   (b) 0.55 sec

(c) 0.7 sec

FIGURE 5.65   Optical microscopic images of Al-brass weld specimens with increase in weld time.

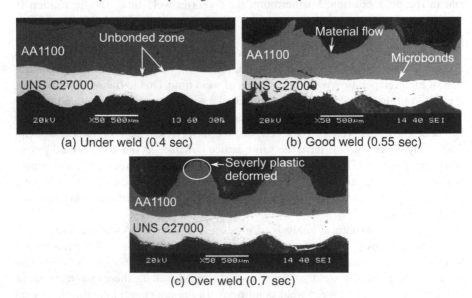

(a) Under weld (0.4 sec)   (b) Good weld (0.55 sec)

(c) Over weld (0.7 sec)

FIGURE 5.66   SEM micrographs of weld cross-sections for Al-brass joints.

FIGURE 5.67 Typical SEM image of top surface of Al sheet under sonotrode tip showing peak and valley regions.

thus the bond density is high in this case. However, higher weld time can produce an oxide layer and cracks, decreasing the weld strength. This condition is termed "over welding."

To consider the good weld condition, where the tensile shear and T-peel failure loads are maximum, the SEM of the same joined sample can be critically examined. The impression marks on the top surface by sonotrode knurls are wholly composed of peaks and valleys, as seen in Figure 5.67. Figure 5.68(a) and (b) depicts the overall fractured surfaces of the Al side and the brass side respectively where the fracture has happened at the periphery of the weld spot. In these fractographs, the previously described four prominent features are also quite visible. From Figure 5.68(a), it can be observed that due to the lower hardness of Al, it is severely deformed, and cracks are found on the surface. Because the brass is harder than Al, it undergoes less deformation, but due to the high plastic deformation of the Al material, the weld zone is almost covered with Al, which is seen as a white patch in Figure 5.68(b). Figure 5.68(c) and (d) represents the magnified images of the Al side and the brass side respectively, where clear vertical lines are not observed due to the high hardness of the brass material. It is obvious that the valley region undergoes more plastic deformation than the peak region. In addition, a squeezed-out eutectic mushy liquid is observed at the edge of the weld spots, confirming the localized melting during the USW process. The temperature generated in this stage is above the recrystallization temperature of both materials. Thus, the flowability of the specimen increases. Figure 5.68(e)–(h) illustrates higher-magnification images of the spots specified on the Al and brass fractured surfaces. In these figures, small island-type weld areas are found along with the dimple and fine dimple fractured regions.

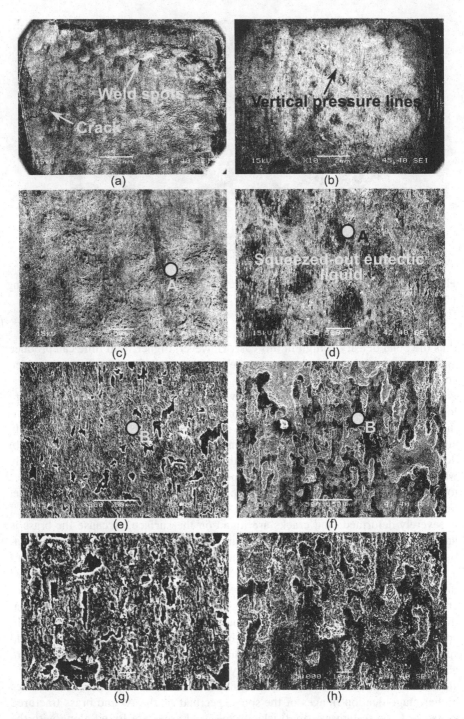

FIGURE 5.68 Overall fractured-surface SEM images of tensile shear failed samples: (a) Al side and (b) brass side; (c) magnified image of (a) showing weld spots; (d) magnified image of (b) indicating squeezed-out eutectic liquid feature; (e) and (f) magnified pictures of spot A denoted in (c) and (d); (g) and (h) higher-magnification images of spot B indicated in (e) and (f).

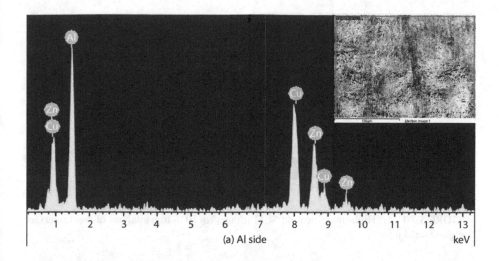

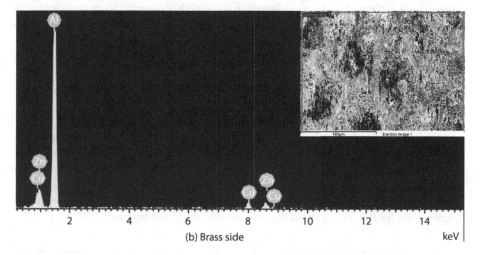

FIGURE 5.69 EDS analysis of fractured surfaces of good weld sample.

Usually, energy dispersive spectroscopy (EDS) analysis is used for the quantitative analysis of elements present on the surfaces. Figure 5.69(a) illustrates EDS analysis conducted on the fractured surfaces of good weld condition (along with SEM images). It shows that a small amount of Cu (i.e., 21.55 wt% and 9.14 wt% of Zn) is present on the fracture surface of the Al side. Likewise, Figure 5.69(b) also confirms that 76.31 wt% of Al is deposited on the brass surface. As Al is a softer material than brass, it is more plastically deformed and sticks to the brass surface, as shown by these results. So, most of the fracture occurs on the low-strength Al side during testing. The SEM micrograph with the backscattered electrons (BSE) can be utilized to reveal the amount of diffusion across the interface for the good weld samples.

A high-magnification SEM figure with EDS mapping is shown in Figure 5.70(a). The line scan result in Figure 5.70(b) reveals a continuous intermetallic compound

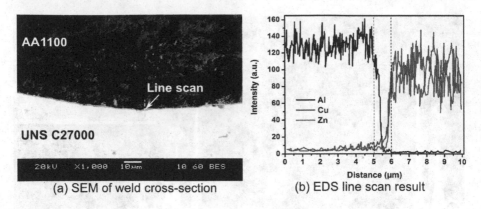

(a) SEM of weld cross-section   (b) EDS line scan result

FIGURE 5.70   (a) High-magnification SEM image; (b) EDS line analysis of weld interface for good weld joint.

layer which is formed at the weld interface through high-temperature generation, as compared with previous Al-Cu welding processes. The diffusion interlayer thickness is about 1 μm, as shown in Figure 5.70(b). As the heterogeneous metals are welded, IMCs such as Al_2Cu, Al_4Cu_9, and Cu_5Zn_8 may form in the interface zone. The chemical composition (wt%) from EDS analysis is 70.26Al–20.12Cu–9.62Zn. This suggests that the dominant elements present in this IMC are probably Al_2Cu and Cu_5Zn_8, according to the Al-Cu phase diagram [9].

XRD analysis confirms the composition of the IMC compounds formed at the faying surface. Figure 5.71(a) and (b) depicts the peak intensities of various compounds with respect to diffraction angle 2θ for good weld sample. The IMCs observed in the weld interface are Al_2Cu, Cu_5Zn_8, and Al_4Cu_9. The other compounds seen are Al on the brass fracture surface and Cu with Zn on the Al fractured surface. The eutectic liquid phase of Cu-Zn proves that Cu_5Zn_8 can probably result in a strong joint and compensate for the formation of the unfavorable Al_2Cu compound.

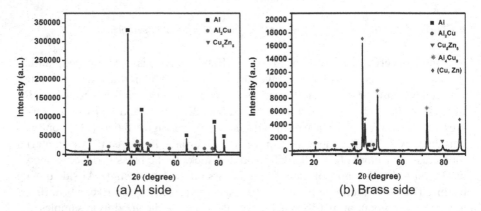

(a) Al side   (b) Brass side

FIGURE 5.71   XRD results showing the presence of various elements on (a) Al fractured surface, (b) brass fractured surface.

REFERENCES

1. Das A, Li D, Williams D, Greenwood D. Joining technologies for automotive battery systems manufacturing. World Electr Veh J 2018;9:22.
2. Kirkpatrick L. Aluminum electrical conductor handbook. Arlington, VA, USA: The Aluminum Association; 1989.
3. Fuhrmann T, Schlegel S, Grossmann S, Hoidis M. Comparison between nickel and silver as coating materials of conductors made of copper or aluminum used in electric power engineering. ICEC 2014; 27th Int. Conf. Electr. Contacts, Dresden, Germany, 17 July 2014, pp. 1–6.
4. Zhao YY, Li D, Zhang YS. Effect of welding energy on interface zone of Al–Cu ultrasonic welded joint. Sci Technol Weld Join 2013;18:354–60. doi:10.1179/13621718 13Y.0000000114.
5. Fujii HT, Endo H, Sato YS, Kokawa H. Interfacial microstructure evolution and weld formation during ultrasonic welding of Al alloy to Cu. Mater Charact 2018;139:233–40.
6. Satpathy MP, Sahoo SK. Microstructural and mechanical performance of ultrasonic spot welded Al–Cu joints for various surface conditions. J Manuf Process 2016;22:108–14.
7. Matsuoka S, Imai H. Direct welding of different metals used ultrasonic vibration. J Mater Process Technol 2009;209:954–60.
8. Wu X, Liu T, Cai W. Microstructure, welding mechanism, and failure of Al/Cu ultrasonic welds. J Manuf Process 2015;20:321–31.
9. Massalski TB, Okamoto H, Subramanian PR, Kacprzak L. Binary alloy phase diagrams. ASM Int Ohio (2nd ed.), 2882; 1990.
10. Liu G, Hu X, Fu Y, Li Y. Microstructure and mechanical properties of ultrasonic welded joint of 1060 aluminum alloy and T2 pure copper. Metals (Basel) 2017;7:361.
11. Fujii HT, Goto Y, Sato YS, Kokawa H. Microstructural evolution in dissimilar joint of Al alloy and Cu during ultrasonic welding. Mater. Sci. Forum 2014;783:2747–52.
12. Suhuddin UFH, Fischer V, Kostka A, dos Santos JF. Microstructure evolution in refill friction stir spot weld of a dissimilar Al–Mg alloy to Zn-coated steel. Sci Technol Weld Join 2017;22:658–65.
13. Nie JF, Zhu YM, Liu JZ, Fang X-Y. Periodic segregation of solute atoms in fully coherent twin boundaries. Science 2013;340:957–60.
14. Panteli A, Chen Y-C, Strong D, Zhang X, Prangnell PB. Optimization of aluminium-to-magnesium ultrasonic spot welding. JOM 2012;64:414–20.
15. Satpathy MP, Sahoo SK. An experimental investigation on joining of aluminium with steel using ultrasonic metal welding. Int J Mechatronics Manuf Syst 2016;9:299–309. doi:10.1504/IJMMS.2016.082861.
16. Li M, Zhu Z, Xiao Q, Zhang Y. Mechanical properties and microstructure evolution of dissimilar Mg and Al alloys welded using ultrasonic spot welding. Mater Res Express 2019;6:86588.
17. Feng M-N, Luo Z. Interface morphology and microstructure of high-power ultrasonic spot welded Mg/Al dissimilar joint. Sci Technol Weld Join 2019;24:63–78.
18. Peng H, Jiang X, Bai X, Li D, Chen D. Microstructure and mechanical properties of ultrasonic spot welded Mg/Al alloy dissimilar joints. Metals (Basel) 2018;8:229.
19. Sato YS, Shiota A, Kokawa H, Okamoto K, Yang Q, Kim C. Effect of interfacial microstructure on lap shear strength of friction stir spot weld of aluminium alloy to magnesium alloy. Sci Technol Weld Join 2010;15:319–24.
20. Chowdhury SH, Chen DL, Bhole SD, Cao X, Wanjara P. Lap shear strength and fatigue behavior of friction stir spot welded dissimilar magnesium-to-aluminum joints with adhesive. Mater Sci Eng A 2013;562:53–60.
21. Feng M-N, Chen Y, Fu D-H, Qu C, Luo Z. Fatigue behaviour and life estimation of Mg/Al ultrasonic spot weld bonding welds. Sci Technol Weld Join 2018;23:487–500.

22. Patel VK, Bhole SD, Chen DL. Microstructure and mechanical properties of dissimilar welded Mg–Al joints by ultrasonic spot welding technique. Sci Technol Weld Join 2012;17:202–6. doi:10.1179/1362171811Y.0000000094.

23. Panteli A, Robson JD, Brough I, Prangnell PB. The effect of high strain rate deformation on intermetallic reaction during ultrasonic welding aluminium to magnesium. Mater Sci Eng A 2012;556:31–42.

24. Shin H-S, de Leon M. Analysis of interface solid-state reaction on dissimilar ultrasonic spot welding of Al-Mg alloys. Met Mater Int 2017;23:554–61.

25. Peng H, Jiang X, Bai X, Li D, Chen D. Microstructure and mechanical properties of ultrasonic spot welded Mg/Al alloy dissimilar joints. Metals (Basel) 2018;8. doi:10.3390/met8040229.

26. Das S, Satpathy MP, Pattanaik A, Routara BC. Experimental investigation on ultrasonic spot welding of aluminum-cupronickel sheets under different parametric conditions. Mater Manuf Process 2019;34:1689–700.

27. Xie Y, Feng M, Cai Y, Luo Z. Ultrasonic spot welding of nickel foam sheet and aluminum solid sheet. Adv Eng Mater 2017;19:1–7. doi:10.1002/adem.201700094.

28. Niu W, Xiao Y, Wan C, Li D, Fu H, He H. Ultrasonic bonding of 2024 Al alloy using Ni-foam/Sn composite solder at ambient temperature. Mater Sci Eng A 2020;771:138663.

29. Ni ZL, Ye FX. Weldability and mechanical properties of ultrasonic welded aluminum to nickel joints. Mater Lett 2016;185:204–7.

30. Mirza FA, Macwan A, Bhole SD, Chen DL, Chen X-G. Microstructure, tensile and fatigue properties of ultrasonic spot welded aluminum to galvanized high-strength-low-alloy and low-carbon steel sheets. Mater Sci Eng A 2017;690:323–36.

31. Mirza FA, Macwan A, Bhole SD, Chen DL, Chen X-G. Effect of welding energy on microstructure and strength of ultrasonic spot welded dissimilar joints of aluminum to steel sheets. Mater Sci Eng A 2016;668:73–85.

32. Macwan A, Kumar A, Chen DL. Ultrasonic spot welded 6111-T4 aluminum alloy to galvanized high-strength low-alloy steel: Microstructure and mechanical properties. Mater Des 2017;113:284–96.

33. Patel VK, Bhole SD, Chen DL. Ultrasonic spot welding of aluminum to high-strength low-alloy steel: microstructure, tensile and fatigue properties. Metall Mater Trans A 2014;45:2055–66.

34. Xu L, Wang L, Chen Y-C, Robson JD, Prangnell PB. Effect of interfacial reaction on the mechanical performance of steel to aluminum dissimilar ultrasonic spot welds. Metall Mater Trans A 2016;47:334–46.

35. Zhu Z, Lee KY, Wang X. Ultrasonic welding of dissimilar metals, AA6061 and Ti6Al4V. Int J Adv Manuf Technol 2012;59:569–74.

36. Zhang CQ, Robson JD, Ciuca O, Prangnell PB. Microstructural characterization and mechanical properties of high power ultrasonic spot welded aluminum alloy AA6111–TiAl6V4 dissimilar joints. Mater Charact 2014;97:83–91. doi:10.1016/j.matchar.2014.09.001.

37. Zhou L, Min J, He WX, Huang YX, Song XG. Effect of welding time on microstructure and mechanical properties of Al-Ti ultrasonic spot welds. J Manuf Process 2018;33:64–73.

38. Wang SQ, Patel VK, Bhole SD, Wen GD, Chen DL. Microstructure and mechanical properties of ultrasonic spot welded Al/Ti alloy joints. Mater Des 2015;78:33–41.

39. Cui X, Fan G, Geng L, Wang Y, Huang L, Peng H-X. Growth kinetics of TiAl3 layer in multi-laminated Ti–(TiB2/Al) composite sheets during annealing treatment. Mater Sci Eng A 2012;539:337–43.

40. Zhang CQ, Robson JD, Prangnell PB. Dissimilar ultrasonic spot welding of aerospace aluminum alloy AA2139 to titanium alloy TiAl6V4. J Mater Process Technol 2016;231:382–8.

6 Thermo-Mechanical Modeling in Ultrasonic Spot Welding of Dissimilar Metal Sheets

Ultrasonic welding (USW) uses high-frequency vibration, usually of 20 kHz or above, with clamping pressure to provide friction-like motion between two metal surfaces. During this motion, the impurities and oxide-layer deformities present on the surfaces disappear and create a suitable environment for metal-to-metal contact. As this is a solid-state welding process, the bonding occurs below the melting point of the base material, instead occurring due to frictional heating, interatomic diffusion, and mechanical interlocking. The most significant advantages of this process are the absence of liquid-solid transformations, the lack of atmospheric contamination, and eco-friendliness, which cannot be claimed for other competing processes like resistance spot welding. In addition, ultrasonic spot welding is the favored technique for joining aluminum and its alloys, as it yields joint strengths comparable to those achieved with other techniques under quasi-static and dynamic loading. The sonotrode is one of the parts of a system that directly touches the upper surface of the specimen; it vibrates parallel to the plane of the weld interface and perpendicular to the axis of clamping force application. Consequently, the vibratory energy is transmitted to the weld spot. Initially, these micro-weld spots are elliptical, but gradually the weld spots increase in size over time to form a continuous weld joint, as shown in Figure 6.1.

6.1 THEORETICAL ASPECTS OF THERMO-MECHANICAL MODELING

The forces at the weld interface during welding, such as shear force (F_{SH}) and static normal force (N), give welders a way to predict unfavorable conditions that will not produce any sound weld. The assumptions made to develop a simple process model of USW are:

i. Only the top and bottom layer/surfaces experience plastic deformation when yield condition is met for the top part.
ii. The compressive stress on the top part is constant and uniform, and is notated as N/S_H or S_{DZ}.
iii. It has been found experimentally that the area in which welding occurs does not go beyond the horn area. This means $S_W \leq S_H$ for all conditions.

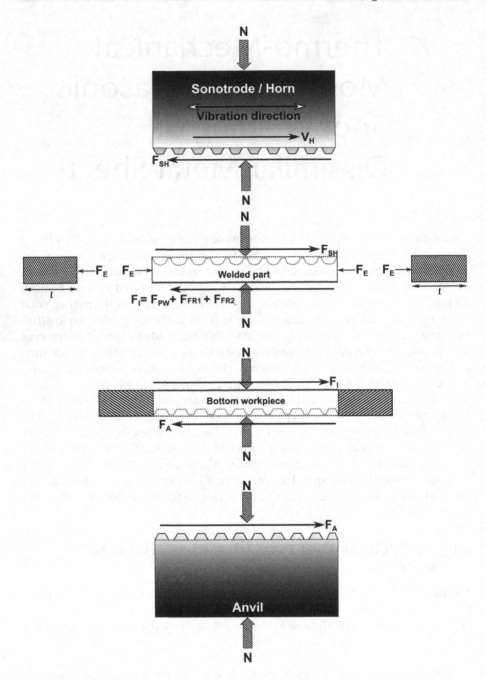

FIGURE 6.1 Detailed analysis of forces acting on individual components during welding.

iv. The elongated part of the top workpiece is treated as the elastic rod excited at the face, which is in contact with the welded part.
v. The bottom workpiece is rigid and fixed as the anvil; hence, the forces acting on the lower surfaces of the bottom workpiece and anvil surfaces are of equal amount and in the opposite direction.

From Figure 6.1, the equilibrium of forces for the top part can be written as

$$m \times \ddot{\xi}_{max} = F_{SH} - F_{PW} - F_{FR1} - F_{FR2} - F_E \tag{6.1}$$

To determine the forces described in Eq. (6.1), *structural analysis* of the welding is carried out.

6.1.1 STRUCTURAL ANALYSIS

When the knurl patterns of the sonotrode and anvil are engaged into a specimen, a complex local stress pattern is produced in the specimen due to the sharp peaks. Thus, elastic-plastic deformation takes place under the peaks and valleys of knurls. However, this complex pattern of stresses will die out at some distance from the sonotrode penetration. Thus, the knurling effect on the stress field will be no longer available. Nevertheless, for the determination of shear force of the horn, a layer of the top workpiece that is just under the sonotrode knurl is taken for investigation. It is assumed that the stress field is uniform in this layer, as shown in Figure 6.2. There are two elements to be considered in this layer. "A" denotes the element under the knurl and "B" represents the element that is present in between the two knurls, where the friction takes place. In the present process model, element A is considered to be the zone where the actual weld takes place.

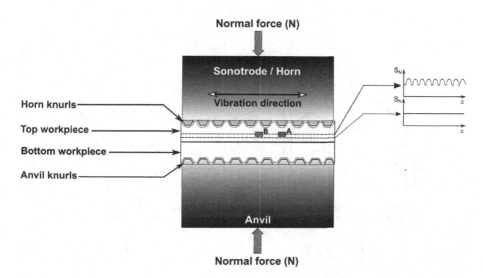

FIGURE 6.2 Stress distribution under the knurls pattern during the USW process.

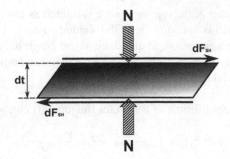

FIGURE 6.3 Small element of the top workpiece under the sonotrode knurl.

However, element A is subjected to plastic deformation during welding because of the synergetic effects of the combined compressive and shearing stresses. The shearing and compressive forces that are acting on this element are presented in Figure 6.3. The limits of the shear stress can be found out by applying principal stress condition and Tresca's yield criteria.

Let's take an inclined plane of a uniform stressed rectangular component, whose thickness is unity. The shear stresses on this plane are equal ($\tau_{xy} = \tau_{yx}$). For better clarification, a three-dimensional (3D) element is used to represent the principal stresses, as shown in Figure 6.4.

Considering the equilibrium of forces in x-direction,

$$\sigma.\cos\theta.b.(1) - \sigma_x.b.\cos\theta.(1) - \tau_{xy}.b.\sin\theta.(1) = 0 \qquad (6.2)$$

$$(\sigma - \sigma_x) = \tau_{xy}.\tan\theta \qquad (6.3)$$

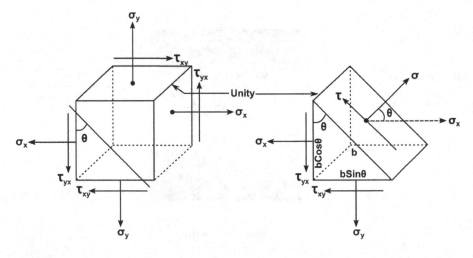

FIGURE 6.4 A 3D element for representation of principal stress.

Similarly considering the equilibrium of forces in y-direction,

$$\sigma.\sin\theta.b.(1) - \sigma_y.b.\sin\theta.(1) - \tau_{yx}.b.\cos\theta.(1) = 0 \tag{6.4}$$

$$(\sigma - \sigma_y) = \tau_{yx}.\cot\theta \tag{6.5}$$

Multiplying Eqs. (6.3) and (6.5)

$$(\sigma - \sigma_x)(\sigma - \sigma_y) = \tau_{xy}.\tau_{yx}.\tan\theta.\cot\theta \tag{6.6}$$

$$(\sigma - \sigma_x)(\sigma - \sigma_y) = \tau_{xy}^2 \tag{6.7}$$

Eq. (6.7) is a quadratic equation and its solutions are

$$\sigma_1 = \left(\frac{\sigma_x + \sigma_y}{2}\right) + \frac{1}{2}\sqrt{(\sigma_x - \sigma_y)^2 + 4.\tau_{xy}^2} \tag{6.8}$$

$$\sigma_2 = \left(\frac{\sigma_x + \sigma_y}{2}\right) - \frac{1}{2}\sqrt{(\sigma_x - \sigma_y)^2 + 4.\tau_{xy}^2} \tag{6.9}$$

In ultrasonic welding, because the forces are acting on the surface vertically, $\sigma_x = 0$. Then, Eqs. (6.8) and (6.9) can be reduced to

$$\sigma_1 = \left(\frac{\sigma_y}{2}\right) + \frac{1}{2}\sqrt{(\sigma_y)^2 + 4.\tau_{xy^2}} \tag{6.10}$$

$$\sigma_2 = \left(\frac{\sigma_y}{2}\right) - \frac{1}{2}\sqrt{(\sigma_y)^2 + 4.\tau_{xy^2}} \tag{6.11}$$

According to the Tresca's yield criteria, the yield will take place when the maximum value of $|\sigma_1 - \sigma_2|, |\sigma_2 - \sigma_3|, |\sigma_3 - \sigma_1| = 2K$ or Y. Thus, from Eqs. (6.10) and (6.11),

$$|\sigma_1 - \sigma_2| = 2\sqrt{\left(\frac{\sigma_y}{2}\right)^2 + \tau_{xy^2}} \tag{6.12}$$

$$|\sigma_2 - \sigma_3| = \frac{\sigma_y}{2} - 2\sqrt{\left(\frac{\sigma_y}{2}\right)^2 + \tau_{xy^2}} \tag{6.13}$$

$$|\sigma_3 - \sigma_1| = -\frac{\sigma_y}{2} - 2\sqrt{\left(\frac{\sigma_y}{2}\right)^2 + \tau_{xy^2}} \tag{6.14}$$

Out of these three equations, Eq. (6.12) will give the maximum value. So

$$|\sigma_1 - \sigma_2| = 2K \text{ or } Y \tag{6.15}$$

$$2\sqrt{\left(\frac{\sigma_y}{2}\right)^2 + \tau_{xy^2}} = Y \tag{6.16}$$

$$\tau_{xy} = \sqrt{\frac{Y^2}{4} - \left(\frac{\sigma_y}{2}\right)^2} \tag{6.17}$$

This τ_{xy} is responsible for the plastic deformation in the volume inside the weld zone. Thus, for the small element shown in Figure 6.3, the shear stress is

$$\tau_S = \frac{dF_{SH}}{dS} \tag{6.18}$$

$$\int dF_{SH} = \int \tau_S . dS \tag{6.19}$$

$$F_{SH} = \sqrt{\frac{Y(T)^2}{4} - \left(\frac{N/S_H}{2}\right)^2} \times S_H \tag{6.20}$$

Cases:

i. If $F_{SH} < \tau_S \times S_H$: Top workpiece will only vibrate and no relative motion is observed between the horn and the workpiece.
ii. If $F_{SH} = \tau_S \times S_H$: Top workpiece will move relative to the motion of the horn, and therefore extrusion and yielding occur in the top part.

It is worth noting that the stresses and plastic deformation occur just below the sonotrode tip when it satisfies the second case. In USW, *sublayer plastic deformation* and *atomic adhesion* are the two primary criteria for joining.

To calculate the shear force at the weld interface, the problem is divided into two parts:

a. Forces related to the dynamics of the top workpiece
b. Interface forces between the workpieces

Both of these problems are analyzed separately in the following sections.

6.1.1.1   Forces Related to the Dynamics of the Top Workpiece

Case I: Both workpieces have the same dimensions as the horn

Figure 6.5 shows the isolated parts from the sonotrode and anvil. The normal and shear forces are acting on both of these components. In this case, at both the

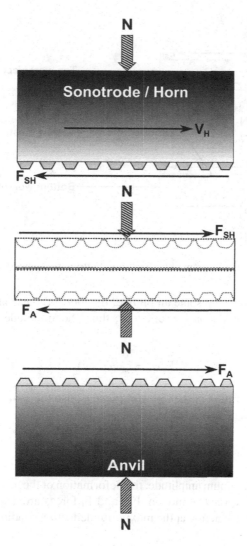

FIGURE 6.5 Free-body diagram of horn, workpieces, and anvil.

top and the bottom workpieces are treated as rigid and have similar dimension as of horn. Due to the rigidness property of the material, elastic deformation due to shear is negligible. It is believed that the movement of the horn will produce an oscillating force on the workpieces. Because these parts have the same dimension as the horn tip, the resonance effect of larger parts is neglected in this case.

During the welding, it has been noticed that initially, friction occurs due to ultrasonic vibration. Thus, some amount of heat is produced, and it leads to the formation of thin film of plastic deformation, as shown in Figure 6.6. In this figure, the two parts are separated from each other to show the interface force F_I.

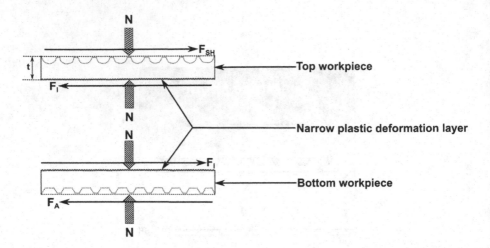

FIGURE 6.6   Free-body diagram of both top and bottom workpieces.

As the horn moves with some amplitude, the top workpiece also moves with the same amplitude. The motion is oscillatory; thus, the amplitude of top part can be written as

$$\xi_T(t) = \xi_h \cos(\omega t) \tag{6.21}$$

Similarly, the acceleration of the top workpiece is also the same as the acceleration of the horn, and can be presented as:

$$\frac{d^2\xi_T}{dt^2} = -\omega^2.\xi_h.\cos(\omega t) \tag{6.22}$$

Therefore, at the maximum amplitude, the deformation of the top workpiece is maximum due to the high relative motion. F_{SH} and F_I forces are also maximum at this point, and the net force acting at the maximum deflection condition is

$$m.\ddot{\xi}_{max} = F_{SH} - F_I \tag{6.23}$$

$$\rho.S_h.t.(\omega^2\xi_h) = F_{SH} - F_I \tag{6.24}$$

In all cases, the bottom part is assumed to be rigid and fixed to the anvil, so there is no relative motion between the anvil and bottom workpiece. Thus, the forces developed here are also the same.

$$F_A = F_I \tag{6.25}$$

Case II: Both workpieces have dimensions different from the horn
The workpieces rarely have the same dimension as the horn tip size. When the top workpiece is extended in both longitudinal and lateral directions, the rigid-part

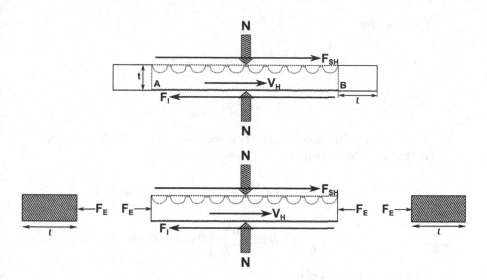

FIGURE 6.7   Free-body diagram of welded part with its forced vibrated elongated part.

assumption about the top surface is no longer valid. This is because the extended part will vibrate elastically, and an anti-resonance condition is developed. Therefore, the top part is divided into two zones: the rigid zone and the elongated zone. These two regions are schematically represented in Figure 6.7.

As previously described, the rigid top part will move according to the movement of the horn. Thus, the elongated part will also get excited according to horn velocity and frequency. Therefore, it is a case of forced vibration, and the elongated part is treated as a rod to find the excitation force (F_E). Here the welding happens only in the rigid zone, so the net force on it will be

$$m.\ddot{\xi}_{max} = F_{SH} - F_I - F_E \tag{6.26}$$

The one-dimensional governing equation for a vibrated top part can be written as

$$\frac{\partial^2 U}{\partial x^2} = \frac{1}{C^2}\frac{\partial^2 U}{\partial t^2} \tag{6.27}$$

The solutions to Eq. (6.27) can be represented as

$$U(x,t) = \left[A\cos\left(\frac{\omega}{C}\right)x + B\sin\left(\frac{\omega}{C}\right)x\right][C\cos\omega t + D\sin\omega t] \tag{6.28}$$

To determine the constants A, B, C, and D, the boundary conditions are applied.

i. When time(t) = 0, initial displacement u(x,0) = 0. Thus, Eq. (6.28) can be written as

$$U(x,0) = \left[A\cos\left(\frac{\omega}{C}\right)x + B\sin\left(\frac{\omega}{C}\right)x\right][C.1] = 0 \tag{6.29}$$

In this case,

$$\left[A \cos\left(\frac{\omega}{C}\right)x + B \sin\left(\frac{\omega}{C}\right)x \right] \neq 0$$

so

$$C = 0 \qquad (6.30)$$

ii. At this point, the velocity also becomes zero, i.e., $\dot{U}(x,0) = 0$. Thus, Eq. (6.28) can be written as

$$\frac{\partial}{\partial t}\left[\left[A \cos\left(\frac{\omega}{C}\right)x + B \sin\left(\frac{\omega}{C}\right)x \right]\left[C \cos\omega t + D \sin\omega t\right]\right] = 0 \qquad (6.31)$$

$$\left[A \cos\left(\frac{\omega}{C}\right)x + B \sin\left(\frac{\omega}{C}\right)x \right] D.\omega.\cos\omega t = 0 \qquad (6.32)$$

In this case, $D.\omega \neq 0$, thus

$$\cos\omega t = 0 \quad \text{and} \quad \omega t = \frac{\pi}{2} \qquad (6.33)$$

After putting the values of Eqs. (6.30) and (6.33) into Eq. (6.28), we get

$$U(x,t) = \left[A \cos\left(\frac{\omega}{C}\right)x + B \sin\left(\frac{\omega}{C}\right)x \right] \times D \qquad (6.34)$$

iii. At the initial starting point, $U(0,t) = \xi_h$. Thus, Eq. (6.35) can be expressed as

$$U(0,t) = [A \times 1 + B \times 0] \times D \qquad (6.35)$$

$$AD = \xi_h \qquad (6.36)$$

iv. The deformation at the end of the rod is also zero. Thus,

$$\frac{\partial U}{\partial x}(l,t) = 0 \qquad (6.37)$$

$$\frac{\partial}{\partial x}\left[A \cos\left(\frac{\omega}{C}\right)x + B \sin\left(\frac{\omega}{C}\right)x \right] \times D = 0 \qquad (6.38)$$

$$\frac{-A\omega}{C} D \sin\left(\frac{\omega}{C}\right)x + BD.\frac{\omega}{C}.\cos\left(\frac{\omega}{C}\right)x = 0 \qquad (6.39)$$

At the end of the rod, $x = 1$. So, Eq. (6.40) becomes

$$\frac{-A\omega}{C} D \sin\left(\frac{\omega}{C}\right)l = -BD.\frac{\omega}{C}.\cos\left(\frac{\omega}{C}\right)l \qquad (6.40)$$

$$BD = \xi_h.\tan\left(\frac{\omega l}{C}\right) \qquad (6.41)$$

After finding out all the constants, the final equation can be written as

$$U = \left[A\cos\left(\frac{\omega}{C}\right)x + B\sin\left(\frac{\omega}{C}\right)x \right] \times D \tag{6.42}$$

$$U = \left[\xi_h.\cos\left(\frac{\omega}{C}\right)x + \xi_h.\tan\left(\frac{\omega l}{C}\right).\sin\left(\frac{\omega}{C}\right)x \right] \tag{6.43}$$

The excitation force for the elongated part can be presented as

$$F_E = \sigma \times S \tag{6.44}$$

$$F_E = S_H \times E \times \frac{\partial U}{\partial x} \tag{6.45}$$

After putting in the values of U from Eq. (6.43), we get

$$F_E = S_H \times E \times \left[\frac{\partial}{\partial x}\left[\xi_h.\cos\left(\frac{\omega}{C}\right)x + \xi_h.\tan\left(\frac{\omega l}{C}\right).\sin\left(\frac{\omega}{C}\right)x \right] \right] \tag{6.46}$$

$$F_E = S_H \times E \times \left[-\xi_h.\left(\frac{\omega}{C}\right).\sin\left(\frac{\omega}{C}\right)x + \xi_h.\left(\frac{\omega}{C}\right).\tan\left(\frac{\omega l}{C}\right).\cos\left(\frac{\omega}{C}\right)x \right] \tag{6.47}$$

The exciting force at the starting point of the elongated part (i.e., x = 0) is given by

$$F_E = S_H \times E \times \xi_h.\left(\frac{\omega}{C}\right).\tan\left(\frac{\omega l}{C}\right) \tag{6.48}$$

Thus, Eq. (6.26) can now be written as

$$m.\ddot{\xi}_{max} = F_{SH} - F_I - F_E \tag{6.49}$$

$$F_{SH} = m.\ddot{\xi}_{max} + F_I + F_E \tag{6.50}$$

$$m.\ddot{\xi}_{max} + F_I + F_E = \sqrt{\frac{Y(T)^2}{4} - \left(\frac{N/S_H}{2}\right)^2} \times S_H \tag{6.51}$$

This equation represents that the yielding and plastic deformation will occur only if the left-hand side (LHS) is equal to the right-hand side (RHS) of the equation. Eq. (6.51) can also be represented as

$$F_I = \sqrt{\frac{Y(T)^2}{4} - \left(\frac{N/S_H}{2}\right)^2} \times S_H - m.\ddot{\xi}_{max} - F_E \tag{6.52}$$

In Eq. (6.52), the first term of the RHS has limited value, and it depends mainly on normal force and temperature. The second term is a constant term for a particular material and geometry. Therefore, if the third term will be high (anti-resonance condition), then F_I will be very low. Hence, welding could not take place; instead, extrusion and yielding will occur.

6.1.1.2 Interface Forces between the Workpieces

Initially, the surfaces of the specimens are covered with contaminants and asperities, and the normal force of the horn is not by itself sufficient to disperse these unwanted materials. Thus, these contaminants still provide resistance to bonding between the surfaces. However, during the welding, these unwanted materials when in contact will plastically deform due to the shearing action of the horn and friction between the sheets. Therefore, substantial heat is developed, which is responsible for lowering the yield strength of the top part. Practically, when the welding process starts, metal-to-metal adhesion occurs in the form of small microbonds, and these areas grow with the increase of time. Finally, the weld area is entirely covered with the deformed material. Consider a small element of the area where metals are in contact with each other and welding occurs, as shown in Figure 6.8.

From Figure 6.8, the yield stress value is given by

$$\tau_S = \frac{dF_{PW}}{dS} \tag{6.53}$$

$$\int dF_{PW} = \int \tau_S \times dS \tag{6.54}$$

From Eq. (6.17), τ_S can be written as

$$F_{PW} = \sqrt{\frac{Y(T)^2}{4} - \left(\frac{N/S_H}{2}\right)^2} \times S_{PW}(t) \tag{6.55}$$

Experimentally, it was found that when the normal force is high, the growth of the weld area is also very high. So, the maximum error may occur. Typically, the weld area is

$$S_{PW} = \frac{S_H}{2} \, or \, \frac{S_{DZ}}{2} \tag{6.56}$$

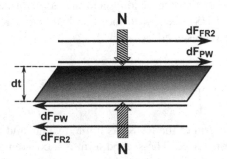

FIGURE 6.8 A small element of weld area subjected to both plastic deformation and friction.

When the total interface force is concerned, then it is divided into two categories:

  i. Force responsible for plastic deformation (F_{PW})
  ii. Total frictional force (F_{FR})

The total frictional force is the net force developed inside the weld zone and also outside the weld zone. Mathematically, it can be expressed as

$$F_I = F_{PW} + F_{FR1} + F_{FR2} \tag{6.57}$$

The frictional force inside the weld zone is given by

$$F_{FR1} = \mu_{S1} \times \sigma_N \times S_{FR1} \tag{6.58}$$

The frictional force outside the weld zone is given by

$$F_{FR2} = \mu_{S2} \times \sigma_N \times S_{FR2} \tag{6.59}$$

Now Eq. (6.1) can be presented as

$$m \times \ddot{\xi}_{max} = F_{SH} - F_{PW} - F_{FR1} - F_{FR2} - F_E$$

Putting the value of Eq. (6.25) in the above equation

$$\rho.S_h.t.(\omega^2 \xi_h) = F_{SH} - F_{PW} - F_{FR1} - F_{FR2} - F_E \tag{6.60}$$

$$\rho.S_h.t.(\omega^2 \xi_h) + F_{PW} + F_{FR1} + F_{FR2} + F_E = \sqrt{\frac{Y(T)^2}{4} - \left(\frac{N/S_H}{2}\right)^2} \times S_H \tag{6.61}$$

In Eq. (6.61), if the LHS is not smaller than the RHS, then yielding will happen during the welding, and consequently tip sticking and extrusion phenomena occur, which are unfavorable conditions of the welding process.

6.1.1.3 Significance of Forces

  F_{SH}: Determines the yielding and extrusion of the top workpiece beneath the horn. Thus, the value of F_{SH} should be less than or equal to shear stress
  F_E: Determines the resonance or anti-resonance of the top workpiece. If it is high, an anti-resonance condition occurs.
  F_{PW}: This force required for welding due to plastic deformation inside the weld spot.
  F_{FR1}: This force is needed for welding due to friction that occurs inside the weld spot.
  F_{FR2}: This force is necessary to determine the friction that occurs outside the weld spot.

6.1.2 THERMAL ANALYSIS

During USW, significant heat is developed at the weld zone, due to plastic deformation as well as friction. Heat significantly affects the material properties. The objective of this exercise is to generate an equation that will give a well-approximated result of acoustical power dissipated to the weld spot. To come up with experimental conditions, it must be realized that the heat developed during the welding process can be divided into two categories:

  i. Heat generation inside the welding spot.
 ii. Heat generation outside the welding spot.

6.1.2.1 Heat Generation Inside the Welding Spot

In this zone, small microbonds are initially developed, and gradually these increase with the increase in weld time. These deformation islands or microbonds occur randomly at the beginning of welding. Because the top workpiece (aluminum) has high thermal conductivity, the temperature will radiate faster from this zone. It is assumed that the areas of all microbonds are the same and also dissipate the same amount of power. Thus, heat generation is given by:

$$\frac{\text{Total Power}(P_{\text{Total}})}{\text{Volume of deformation zone}(\vartheta_{DZ})} = \frac{\sum P_i}{\vartheta_{DZ}} \tag{6.62}$$

$$\frac{P_{Total}}{\upsilon_{DZ}} = \frac{\int dp}{\upsilon_{DZ}} \tag{6.63}$$

Firstly, it is assumed that the heat generated in the weld zone is due to the plastic deformation of the material. A layer under shearing action is shown in Figure 6.9.

The work done by this deformed volume for a particular period is equal to the shear angle change within that time. Mathematically,

$$\frac{d\Delta w}{\Delta T_i} \times \frac{1}{d\upsilon} = \tau_S \times \frac{\Delta \xi_T}{dT_i} \times \frac{1}{\Delta T_i} \tag{6.64}$$

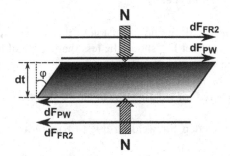

FIGURE 6.9 Small element under shear strain for heat generation.

In Eq. (6.65),

$$\frac{\Delta w}{\Delta T_i} = Power(P)$$

$$\frac{\Delta \xi_T}{\Delta T_i} = Velocity(V_{avg})$$

(6.65)

Thus, Eq. (6.65) can be expressed as

$$\frac{dP}{d\upsilon} = \tau_S \times \frac{V_{avg}}{dT_i}$$

(6.66)

The ultrasonic wave is believed to be sinusoidal, and it passes through the weld; thus, the total time is divided into four sections, as shown in Figure 6.10.

Hence, Eq. (6.65) can be written as

$$V_{avg} = \frac{\displaystyle\int_0^{T_i} \Delta \xi_T}{\displaystyle\int_0^{T_i} \Delta T_i}$$

(6.67)

From Eq. (6.21), ξ_T can be written as

$$V_{avg} = \frac{1}{(T_i - 0)} \times \int_0^{T_i} \left| \xi_h \times \omega \times \cos \omega T_i \right| . dT_i$$

(6.68)

$$V_{avg} = \frac{1}{(T_i)} \times \xi_h \times \omega \times \frac{1}{\omega} \times [[\sin \omega T_i]_0^{T_i/4} - [\sin \omega T_i]_{T_i/4}^{T_i/2} - [\sin \omega T_i]_{T_i/2}^{3T_i/4} + [\sin \omega T_i]_{3T_i/4}^{T_i}]$$

(6.69)

$$V_{avg} = 4 \times \xi_h \times f_{PW}$$

(6.70)

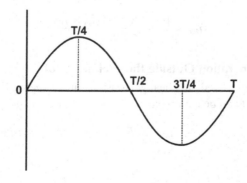

FIGURE 6.10 Sinusoidal ultrasonic wave.

Putting the values of Eq. (6.70) into Eq. (6.66), we get

$$\frac{dP}{d\upsilon} = \tau_s \times \frac{4 \times \xi_h \times f_{PW}}{dT_i} \tag{6.71}$$

$$\frac{dP}{dS \times dx} = \tau_s \times \frac{4 \times \xi_h \times f_{PW}}{dx} \tag{6.72}$$

$$\int dP = \int \tau_s \times 4 \times \xi_h \times f_{PW} \times dS \tag{6.73}$$

Putting the value of Eq. (6.73) into Eq. (6.63), we get

$$\frac{P_{Total}}{\upsilon_{DZ}} = \frac{\int dP}{\upsilon_{DZ}} \tag{6.74}$$

$$\frac{P_{Total}}{\upsilon_{DZ}} = \frac{\int \tau_s \times 4 \times \xi_h \times f_{PW} \times dS}{\upsilon_{DZ}} \tag{6.75}$$

$$\frac{P_{Total}}{S_{DZ} \times dt} = \frac{\int \tau_s \times 4 \times \xi_h \times f_{PW} \times dS}{S_{DZ} \times dt} \tag{6.76}$$

$$\dot{q}_{PW} = \frac{P_{Total}}{S_{DZ}} = \frac{4 \times \xi_h \times f_{PW} \times F_{PW}(t)}{S_{DZ} \times dt} \tag{6.77}$$

Eq. (6.77) represents the amount of acoustic power that is developed due to plastic deformation, and this is the heat produced at the weld zone.

Similarly, another source of heat inside the weld spot is friction. Hence, it can be presented as

$$\frac{P_{FR1}}{S_{FR1}} = \frac{F_{FR1} \times V_{avg}}{S_{FR1}} \tag{6.78}$$

$$\dot{q}_{FR1} = \frac{\mu_{S1} \times N \times 4 \times \xi_h \times f_{PW}}{S_{FR1}} \tag{6.79}$$

6.1.2.2   Heat Generation Outside the Welding Spot

As in the case just discussed, any heat generation outside the weld spot is due only to friction. So, the heat can be expressed as

$$\frac{P_{FR2}}{S_{FR2}} = \frac{F_{FR2} \times V_{avg}}{S_{FR2}} \tag{6.80}$$

$$\dot{q}_{FR2} = \frac{\mu_{S2} \times N \times 4 \times \xi_h \times f_{PW}}{S_{FR2}} \tag{6.81}$$

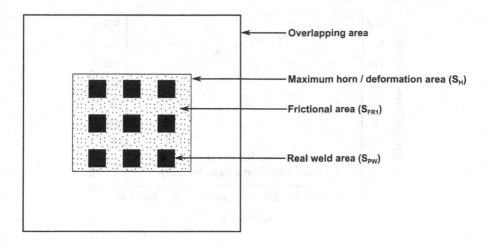

FIGURE 6.11   Schematic diagram of different time areas [1].

In Eq. (6.81), S_{FR2} can be calculated as

$$S_{FR2} = S_H - S_{PW} \tag{6.82}$$

Schematically, this area can be observed in Figure 6.11.

6.1.3   ACOUSTIC SOFTENING ANALYSIS

Acoustic softening is the second reason for material softening, in which the static stress of the material is significantly reduced under the influence of ultrasonic energy. It was first observed by Langenecker [2]. It differs from material to material and mainly depends on acoustic impedance, melting point, Young's modulus, and hardenability of the material. One should consider acoustic softening along with thermo-mechanical effects for the analysis of USW mechanics. Without knowledge of it, the stress field and plastic deformation cannot be accurately modeled. Thus, we require a method to quantify acoustic softening, to characterize it under the broad range of process parameters, geometries, and materials.

6.1.3.1   Preparation of Material Model

As an illustration, in the subsequent analysis, AA 1100 aluminum alloy is used as one of the weld materials, which is usually placed on the top during the welding process. Hence, a material model is necessary to find the relationship between acoustic softening, thermal softening, and strain hardening. Hockett [3] observed the strain-hardening behavior of AA1100 material in relation to thermal softening.

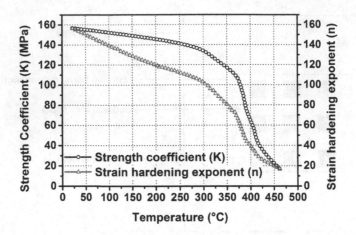

FIGURE 6.12   Variation in power law constants concerning temperature for AA1100.

He presented the stress-strain relationship at a particular temperature in the plastic region by using the power-law equation shown here:

$$\sigma_y = K_S \times \varepsilon_P^n \tag{6.83}$$

In this equation, the constants such as K_S and n at different temperatures were derived by Hockett [3]. The intermediate values are calculated by using linear interpolation, as represented in Figure 6.12.

In the present model, the power-law model given by Hockett is marginally modified by adding an acoustic softening parameter (β). The range of this newly added term varies from zero to one. If $\beta = 1$, then it is believed that the acoustic softening has no impact on the material deformation and the material is deformed normally. But when $\beta = 0$, it implies that the ultrasonic energy deforms the material sufficiently. The following shows the power-law equation with the acoustic softening parameter:

$$\sigma_y = \beta \times K_S \times \varepsilon_P^n \tag{6.84}$$

It is assumed that the effect of parameter n on the yield strength of the material is negligible and that K_S is the critical term that most affects the yield strength (σ_y) of the material. Thus, the acoustic softening parameter linearly affects the K_S term.

6.1.3.2   Acoustic Softening Term Calculation

To quantify the acoustic softening in the sheets during ultrasonic metal welding (USMW) process, the β is decreased from one to zero until the error between the modeling result (i.e., increase in sheet width) and the experimental result is minimized. For clear understanding, a contour plot of a modeled sheet deformation is shown in Figure 6.13.

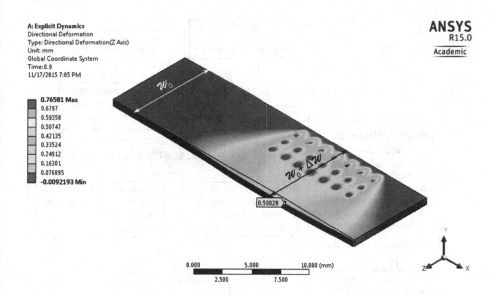

FIGURE 6.13 Contour plot of a sheet with initial sheet width and deformed sheet dimension.

A flow chart is shown in Figure 6.14, summarizing all the steps to be followed in the thermo-mechanical and acoustic softening analysis.

The other approach to representing the material softening in USW is to regard it as the synergetic effect of acoustic softening and thermal softening. The amount of material softening is decided by the decrement in the yield strength of the material. The yield strength of the material (σ_s) during the USW process can be expressed as

$$\sigma_S = \sigma_T \times \alpha_{US} \tag{6.85}$$

where σ_T represents the temperature-dependent yield stress of the material and α_{US} is the ultrasonic softening rate, which decreases with the decrement in the yield stress of materials. This ultrasonic softening rate is related to vibration amplitude and frequency during the welding process [4]. It is represented as

$$\alpha_{US} = (1 - d \times I_{US})^a \tag{6.86}$$

where d and a are the ultrasonic softening related parameters and I_{US} is the ultrasonic intensity related to ultrasonic vibration amplitude and frequency. I_{US} can be figured out by

$$I_{US} = \frac{\rho}{2} \times v \times (2\pi \times f \times \xi)^2 \tag{6.87}$$

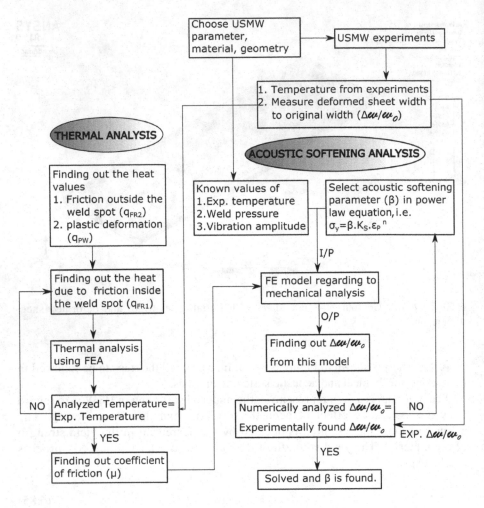

FIGURE 6.14 Flow chart of the thermo-mechanical model for quantifying acoustic softening.

where ρ and v are the density of the weld materials and velocity of the sound waves in these materials, and f and ξ are the vibration frequency and amplitude, respectively.

The a and d constant parameters in Eq. (6.86) can be established using the inverse modeling technique. Huang et al [5]. and Hung and Hung [6] utilized this method to find out these parameters for Al and Cu materials, respectively. It can be defined as follows:

The ultrasonic softening rate for Al:

$$\alpha_{Al} = \left[1 - 0.1056 \times \left(\frac{f \times \xi_{Al}}{0.112} \right)^2 \right]^2 \tag{6.88}$$

Likewise, the ultrasonic softening rate for Cu:

$$\alpha_{Cu} = \left[1 - 0.2864 \times \left(\frac{f \times \xi_{Al}}{0.170}\right)^2\right]^5 \tag{6.89}$$

Meanwhile, earlier studies on acoustic softening by Langenecker [2], Siddiq [7], and El Sayed [8] reveal that the stress reduction in the material is proportional to the ultrasonic energy. The amount of material softening can be achieved by introducing a softening term based on the ultrasonic energy density. This leads to a modification in the Johnson-Cook model to counter the acoustic softening effect (ASE) during the USW process. This ASE has an independent role in accelerating the material deformation. The independent softening term can be expressed as

$$\xi_{Ultrasonic} = \exp(n_2, \rho_{Ultra}) \tag{6.90}$$

The modified constitutive Johnson-Cook model incorporating the ASE term can be expressed as

$$\sigma^0 = \xi_{Ultrasonic}\left[A + B\left(\bar{\varepsilon}^{pl}\right)^n\right]\left[1 + C\ln\left(1 + \frac{\dot{\varepsilon}}{\dot{\varepsilon}_0}\right)\right]\left(1 - \hat{\theta}^m\right) \times \exp\left(n_2, \rho_{Ultra}\right) \tag{6.91}$$

All the components of Eq. (6.91) can be obtained through tensile tests and ultrasonic vibration-assisted compression tests at elevated temperature. Moreover, a comparative model can be constructed with and without considering the ASE to demonstrate the effect of acoustic softening on this thermo-mechanical process quantitatively.

6.2 APPLICATION OF NUMERICAL MODELING

As an illustration, thermo-mechanical FE modeling is described to analyze the USW of Al-Cu joints incorporated with commercial software such as ANSYS® and ABAQUS®. Several steps are followed in this process, and these are discussed later in this section:

- Finding out the compressive stress distribution during the delay time for predicting its uniformity beneath the sonotrode knurl.
- Predicting contact stress on the bottom sheet during the delay time, showing the thickness of the sheets up to which they can be weldable.
- Exploring the temperature distribution at the sonotrode-top part weld interface with respect to weld time.

6.2.1 COMPRESSIVE STRESS ANALYSIS UNDER SONOTRODE KNURLS DURING THE DELAY TIME

We consider 0.7-mm-thick aluminum and 0.4-mm-thick copper sheets for demonstrating the FE analysis during the USW process. Initially, at the starting time of the weld, the sonotrode knurls press the aluminum and copper sheets firmly without any

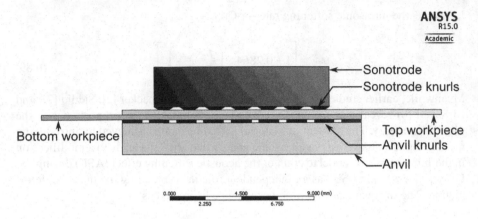

FIGURE 6.15   Schematic model used for thermo-mechanical analysis.

vibration until the delay time is over. For the analysis point of view, a constant 0.2 MPa clamping pressure is utilized. This penetration is necessary to provide adequate friction during the welding process. Thus, to find out the compressive stress exerted by the sonotrode on the top sheet, the FEA program ANSYS® is used. Meantime, the maximum depth of indentation depends on the height of the sonotrode knurls, and it is considered to be 0.2 mm. Figure 6.15 depicts the schematic model of a total system which is used for further analysis. Due to less welding time, relatively less thermally conductive sonotrode and anvil material, and for simplification of the 3D problem, only half of the total system is taken into consideration. Actual material properties values of common Al and Cu sheets are used for the simulation.

To do this analysis, the static structural module of ANSYS® is used with functional boundary conditions. As the correctness of the FEA solution depends on the mesh element size and its types, hexahedral meshing is selected for the sonotrode and two sheets and tetrahedral meshing for the anvil. The number of mesh elements varies from one part to another, and here the finest mesh size is chosen for the weld interface. The material properties included in the model are Young's modulus, Poisson's ratio, specific heat, thermal conductivity, and density, as presented in Table 6.1. Figure 6.16 illustrates the meshing for the whole system. The boundaries of this model are considered to be frictionless rigid walls, so as to find out the stress distribution beneath the sonotrode knurls.

6.2.2   HEAT DEVELOPED DUE TO PLASTIC DEFORMATION

The experimental parameters used in the numerical analysis are the temperature-dependent weld area at various surface conditions and the normal static load exerted by the sonotrode. As USW is a complex phenomenon, the boundary conditions for Eq. (6.77) are also quite complicated. These boundary conditions are based on the

TABLE 6.1
Physical, Mechanical, and Thermal Properties of Weld Materials

Properties	Unit	Materials			
		AA1100	UNS C10100	UNS C27000	SS 304
Density	kg/m³	2710	8940	8470	8000
Young's modulus	Pa	68.9E9	115E9	105E9	193E9
Poisson's ratio		0.33	0.31	0.3	0.29
Ultimate tensile strength	Pa	135.5E6	302.1E6	420.4E6	858.1E6
Yield strength	Pa	115.2E6	251.5E6	336.1E6	649.4E6
Co-efficient of thermal expansion	°C⁻¹	23.6E-6	17E-6	20.3E-6	17.3E-6
Thermal conductivity	W/m°C	220	391	116	16.2
Specific heat	J/kg°C	904	385	380	500
Melting point	°C	660	1085	930	1455

temperature-dependent yield strength of the material, normal static load applied on the sheet, and average velocity of the sonotrode at which deformation takes place. The temperature-dependent yield strength is illustrated in Figure 6.17.

For simplification of the problem, linear approximation can be used, and the corresponding temperature-dependent yield strength can be written as

$$Y(T) = (-0.2712T + 115) \times 10^6 \, Pa \tag{6.92}$$

ANSYS
R15.0
Academic

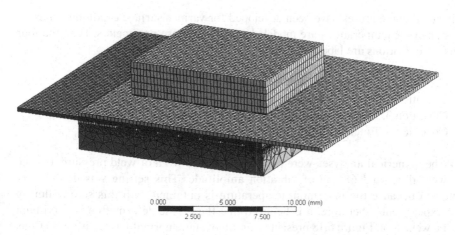

0.000   5.000   10.000 (mm)

2.500   7.500

FIGURE 6.16 Various mesh elements used in the proposed FE model.

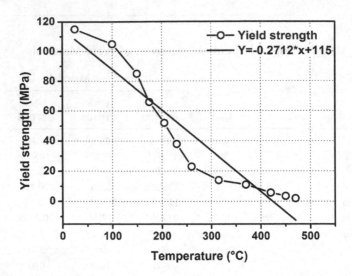

FIGURE 6.17 Linear approximation of yield strength with temperature.

The average temperature-dependent yield strength can be derived by taking the integration of Eq. (6.92) with a specific range of temperatures. In this analysis, the $Y(T)$ can be written as:

$$Y(T) = \int_{25}^{470} \frac{(-0.2712T + 115)dT}{\Delta T}$$

$$Y(T) = 47.9 \times 10^6 \, Pa \tag{6.93}$$

Different temperatures have been developed for various surface conditions; thus, in this analysis, temperatures are modeled according those parameters. Here the four surface conditions are labeled as:

Condition 1: Lubricating surface
Condition 2: Normally polished surface
Condition 3: Electrolytically polished surface
Condition 4: Emery polished surface

All the numerical analyses were performed at 0.38 MPa of weld pressure, 0.9 sec of weld time, and 68 μm of vibration amplitude. This setting was deliberately selected because the maximum temperature is obtained with this setting during the experiment. Therefore, it is believed that the plastic deformation is maximum at the weld spot. Figure 6.18 presents a graph of static normal force with weld time. It can be observed that the static normal load increases with an increase in weld

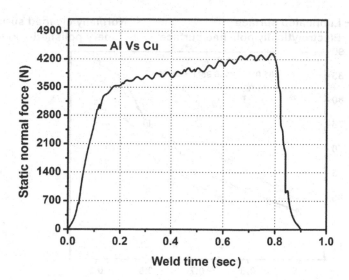

FIGURE 6.18   Normal static load variation with respect to weld time.

time and that the maximum value of 4356 N is obtained at 0.9 sec. As the weld pressure remains the same for all conditions, this normal static load value is also the same for all.

It is evident that the strength of the weld increases with an increase in weld time up to a particular value. Meantime, the generation of temperature also increases with time, as does the weld area. Thus, the plastically deformed weld area at the weld spot is found out from the polynomial curve fitting for each surface condition. The curve-fitting equations of weld areas for each surface condition are shown in Figure 6.19. At the end of the weld time, the weld areas are assumed to be fully developed, and experimentally these are obtained as 88.8 mm², 85.1 mm², 79.9 mm², and 76.6 mm² for conditions 1, 2, 3, and 4, respectively.

The welding is performed with a constant vibration amplitude of 68 μm. At the initial stage of welding, the vibration amplitude increases very quickly and then attains a constant value. This constant value is used to find out the average velocities of the sonotrode. However, the average velocity of the sonotrode is different for different surface conditions, as the coefficient of friction varies. Thus, in the case of maximum surface roughness, the average velocity will be lowest. As the sonotrode moves in a linear direction, the concept of band brake comes into the picture. According to this concept, the frictional work done can be found as:

$$-\mu mgx = -\frac{1}{2}mV_0^2 \text{(-ve sign due to retradation)}$$

$$x = \frac{V_0^2}{2\mu g} \qquad\qquad (6.94)$$

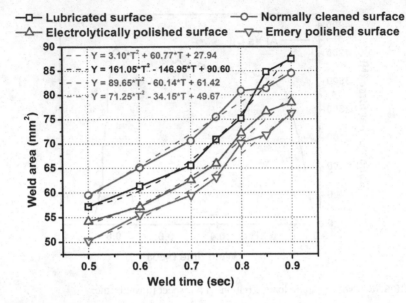

FIGURE 6.19   Weld area variation vs. weld time for various surface conditions with curve-fitting equations.

where x is the stopping distance, m is the mass of the body, and V_0 denotes the velocity of the horn. Meantime, the sonotrode also oscillates in simple harmonic motion (SHM). Thus, the linear distance covered by it is given by

$$x = \xi \mathrm{Sin}\omega t$$

$$\xi \mathrm{Sin}\omega t = \frac{V_0^2}{2\,\mu g}$$

$$\frac{\xi_1}{\xi_2} = \frac{\mu_2}{\mu_1} \tag{6.95}$$

In the lubricating case, $\xi_1 = 68\ \mu m$, $\mu_2 = 0.31$, and $\mu_1 = 0.015$. Thus, for condition 1:

$$\frac{68}{\xi_2} = \frac{0.31}{0.015}$$

$$\xi_2 = 3.29\ \mu m \tag{6.96}$$

Therefore, the average velocity of the sonotrode can be determined by using Eq. (6.70):

$$V_{avg} = 4 \times \xi_h \times f_{PW}$$

$$V_{avg} = 4 \times 3.29 \times 10^{-6} \times 20000$$

$$V_{avg} = 0.26\ m\,/\,s \tag{6.97}$$

Similarly, for condition 2, 3, and 4, the average velocities obtained are 0.20 m/sec, 0.15 m/sec, and 0.11 m/sec, respectively.

After finding out all these terms, the heat due to plastic welding can be found out for four different surface conditions. This expression gives the heat as a function of temperature, normal static load, weld area, and average velocity of the horn. Thus, Eq. (6.77) can be written as:

For condition 1:

$$q_{PW} = \sqrt{\frac{Y(T)^2}{4} - \left(\frac{N/S_H}{2}\right)^2} \times S_{PW}(t) \times 4 \times \xi_2 \times f_{PW}$$

$$q_{PW} = \sqrt{\left(\frac{47.88 \times 10^6}{2}\right)^2 - \left(\frac{4356 \times 10^6}{2 \times 99}\right)^2} \times 88.8 \times 10^{-6} \times 0.26$$

$$q_{PW} = 220.58 \text{ watt} \tag{6.98}$$

Similarly, for conditions 2, 3, and 4, the heat generated due to plastic welding is determined to be 163.92 watts, 109.90 watts, and 78.70 watts, respectively. Furthermore, one can determine the generation of temperature due to heat by applying the specific heat concept. The relationship can be written as:

For condition 1:

$$q_{PW} \times t = C_P \times m \times \Delta T$$

$$q_{PW} \times t = C_P \times (\rho \times V) \times (T_{max1} - T_{room})$$

$$220.58 \times 0.9 = 904 \times (2710 \times 20 \times 80 \times 0.7 \times 10^{-9}) \times (T_{max1} - 23)$$

$$T_{max1} = 95.48°C \tag{6.99}$$

where t = weld time in sec, C_p = specific heat of the aluminum sheet, ρ = density of the material, V = volume of the material, T_{max} = maximum temperature generation during welding, and T_{room} = room temperature.

Similarly, 76.86°C, 59.11°C, and 48.86°C temperatures are developed during the welding process at the plastic deformation zone.

6.2.3 HEAT DEVELOPED DUE TO FRICTION OUTSIDE THE PLASTIC DEFORMATION ZONE

The heat due to friction outside the welding spot can be determined using Eq. (6.81). With the static coefficient of friction, normal static load, and average velocity of the sonotrode, one can find out the heat for four surface conditions. The static coefficient

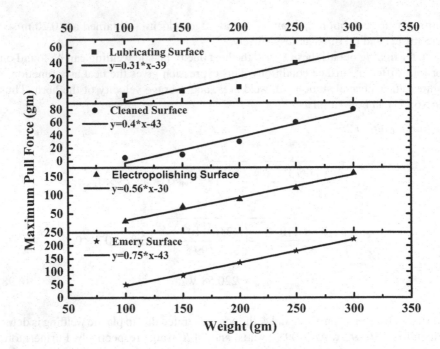

FIGURE 6.20 Coefficient of friction for Al-Cu weld materials with various surface conditions.

of friction can be obtained from the curve-fitting equations used for different surface conditions shown in Figure 6.20.

Likewise, the normal static load can be drawn from Figure 6.18. After finding out all the values, the heat developed due to friction outside the plastic welding zone can be written as:

For condition 1:

$$q_{FR2} = \mu_{S2} \times N \times 4 \times \xi_2 \times f_{PW}$$

$$q_{FR2} = 0.31 \times 4356 \times 0.26$$

$$q_{FR2} = 355.45 \text{ watt} \tag{6.100}$$

For conditions 2, 3, and 4, this value remains the same after calculation. Moreover, the temperature generated due to friction can be found by using the same concept used previously. It can be written as:

For condition 1:

$$q_{FR2} \times t = C_P \times m \times \Delta T$$

$$q_{FR2} \times t = C_P \times (\rho \times V) \times (T_{\max 2} - T_{room})$$

$$355.45 \times 0.9 = 904 \times (2710 \times 20 \times 80 \times 0.7 \times 10^{-9}) \times (T_{\max 2} - 23)$$

$$T_{\max 2} = 139.79°C \tag{6.101}$$

For conditions 2, 3, and 4, similar temperature values are obtained.

6.2.4 Heat Developed Due to Friction Inside the Plastic Deformation Zone

According to the principles of USW, metallurgical bonding occurs between the two contact surfaces with the application of normal force and shear vibration of the sonotrode. Thus, the contact stresses between the parts spread out as weld time increases, and the rate of spread depends on the thickness of the sheets and the anvil area. In observing the fractured weld surfaces of sheets, one can clearly see the plastically deformed areas, abraded areas outside the weld spot, and the field inside the weld spot where welding has not occurred. This strongly indicates that before the plastic deformation occurred, friction took place between the sheets. The unwelded areas inside the weld spot signify that at a shorter weld time, the contaminants and asperities present over the weld surface were not entirely removed by shear vibration. Thus, in this type of zone, only friction takes place, which is similar to the friction that occurred outside the weld spot. It is tough to find out which frictional forces acted on this zone. However, heat developed in this area has a significant effect in lowering the yield strength of the material. This phenomenon is called *acoustic softening*. Here, an attempt has been made to consider the heat developed in this zone and make it one of the parameters in F_E analysis of temperature. To determine the heat generated in this region, first one must find out the corresponding temperature, by using the experimental data.

For condition 1:

Experimental temperature $(T_{total}) = 375.37°C$
Temperature due to plastic deformation $(T_{max1}) = 95.48°C$
Temperature due to friction outside the plastic deformation zone $(T_{max2}) = 139.79°C$
Temperature due to friction inside the plastic deformation zone

$$T_{max3} = 375.37 - (95.48 + 139.79) = 140.10°C \tag{6.102}$$

Now, the temperature evolved inside the weld spot for conditions 2, 3, and 4 are obtained as 133.70°C, 120.19°C, and 106.43°C, respectively.

To find out the heat due to the friction that happened inside the weld spot, the specific heat concept can be used. The relationship can be written as:

$$q_{FR1} \times t = C_P \times m \times \Delta T$$

$$q_{FR1} \times 0.9 = 904 \times (2710 \times 20 \times 80 \times 0.7 \times 10^{-9}) \times (T_{max3} - 23)$$

$$q_{FR1} = 356.39 \text{ watt} \tag{6.103}$$

This q_{FR1} varies from one surface condition to another. Thus, 336.91 watts, 295.79 watts, and 253.92 watts are obtained for conditions 2, 3, and 4, respectively. These heat

TABLE 6.2
Overall Summary of Results Obtained from USMW Mechanics

	T_{total} in °C	$Y(T)$ in MPa	N	S_H in mm²	S_{PW} (T) in mm²	V_{avg} in m/sec	μ_{S2}	q_{PW} in watts	Q_{FR2} in watts	T_{max1} in °C	T_{max2} in °C	T_{max3} in °C	Q_{FR1} in watts	μ_{S1}
Condition 1	375.37	47.9	4356	9.90	8.88	0.26	0.31	220.58	355.45	95.48	139.79	140.10	356.39	0.31
Condition 2	350.35	47.9	4356	9.90	8.51	0.20	0.4	163.92	355.44	76.86	139.79	133.70	336.91	0.38
Condition 3	319.09	47.9	4356	9.90	7.99	0.15	0.56	109.90	355.45	59.11	139.79	120.19	295.79	0.47
Condition 4	295.08	47.9	4356	9.90	7.66	0.11	0.75	78.70	355.45	48.86	139.79	106.43	253.92	0.54

values can be further used to find out the coefficient of friction occurring inside the plastic deformation zone. Thus, it can be written as:

For condition 1:

$$q_{FR1} = \mu_{S1} \times N \times 4 \times \xi_2 \times f_{PW}$$

$$356.39 = \mu_{S1} \times 4356 \times 0.26$$

$$\mu_{S1} = 0.31 \tag{6.104}$$

Similarly, for conditions 2, 3, and 4, the friction values are observed as 0.38, 0.47, and 0.54, respectively. These coefficient of friction values are crucial for finding out the acoustic softening parameter, which is described in the next section. The overall summary of all these calculated values is given in Table 6.2.

All the heat values from Eqs. (6.98), (6.100), and (6.103) are known, and these can be applied directly to the FE model. Furthermore, the boundary conditions are also applied to the model, as shown in Figure 6.21. In this figure, q_{conv} represents the heat lost to the surroundings due to convection, and the overall heat transfer coefficient (h) is taken as 5 J/m²°C. The heat due to plastic deformation (q_{PW}) and friction inside the weld zone (q_{FR1}) are applied to the areas where sonotrode tips touch the top workpiece. Likewise, the areas where the tips are not in contact are treated as having the friction outside the weld spot, and q_{FR2} is applied in these regions. Convection conditions occur in those areas of specimens which are not in contact with either sonotrode or anvil. The reference temperature is taken as 23°C, and the transient thermal analysis has been carried out with a time step of 0.1 sec.

Any overview of thermo-mechanical modeling of a friction-based solid-state welding technique such as USW must confront the following issues:

    i. Interface boundary conditions at specific locations.

    ii. Specification of heat input sources during the USW process.

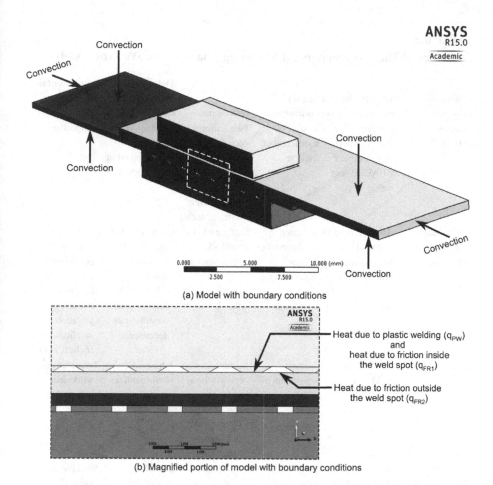

(a) Model with boundary conditions

(b) Magnified portion of model with boundary conditions

FIGURE 6.21   Boundary conditions applied to thermo-mechanical model [1].

iii. Choice of suitable constitutive material model with the addition of ultra-sonic softening.

iv. Validation and calibration of heat input, temperature, and deformation of weldments.

The evolution of this modeling process has been addressed by various researchers over time, and a summary of their research outcomes regarding these issues is given in Table 6.3.

Summarizing the previous studies, most thermo-mechanical modeling primarily focuses on the prediction of interface temperature generation and deformation as a function of process parameter combinations and geometries of sonotrode and anvil knurls. Meanwhile, it is inferred from the USW principle that in the early stages of asperity contact, the heat input is mainly provided by the Coulomb friction with a coefficient of friction. This coefficient of friction can be considered as constant,

TABLE 6.3

Summary of Thermo-mechanical Modeling Analysis in USW Process [9]

Authors/ references	Material and geometry	Contact conditions	Heat input	Material constitutive data	Experimental validation/ calibration
Elangovan et al. [10, 11]	Al sheet; Al2O3; Al-Al and dissimilar Al-Al2O3 spot welds	Constant coefficient of friction	Two uniform, constant heat fluxes (one for friction, one for deformation, applied over different areas)	Constant yield stress, independent of temperature	Temperature
Siddiq and Ghassemieh [12, 13]	Al 6061 foil and sheet; Al 3003 foil and sheet; seam weld	Calibrated coefficient of friction, dependent on process conditions	Surface friction and bulk plasticity	Cyclic plasticity model with thermal and acoustic (ultrasonic) softening, and no strain-rate dependence	No direct validation; only empirical correlations between fracture energy, vibration amplitude, and friction work
Kim et al. [14]	Al 5754; spot weld	Constant coefficient of friction	Surface friction (proportional to friction force and slip velocity) and consequent bulk plasticity	Johnson-Cook deformation model	Temperature, workpiece velocities
Doumanidis and Gao [15, 16]	Al 1100 foil and Al 6061 sheet; US consolidation	Analytical model and calibrated coefficient of friction	Mechanical analysis only, as precursor to subsequent thermal simulation	Constant (ambient) temperature; bilinear elastic-plastic model with kinematic hardening	Workpiece velocities, temperature–time histories
Zhang and Li [17–19]	Al 3003 foil; ultrasonic consolidation	Two cases: constant and temperature-dependent coefficients of friction, calibrated via separate experiments	Two cases: friction only (from shear stress and sliding distance), and friction plus plastic heating	Two cases: constant (warm) yield-stress, and temperature-dependent yield stress	No temperature validation

TABLE 6.3 *(Continued)*
Summary of Thermo-mechanical Modeling Analysis in USW Process [9]

Authors/ references	Material and geometry	Contact conditions	Heat input	Material constitutive data	Experimental validation/ calibration
Lee et al. [20] Lee and Cai [21]	Al and Cu foil/ sheet; multi-sheet/ dissimilar spot welds	Constant coefficient of friction, adjusted by material and joint configuration	Two cases: friction only, and friction plus plastic heating	Cyclic plasticity combining isotropic and kinematic hardening, with temperature dependency	No direct validation; only weld surface profiles, and indicative final temperatures from other work
Chen et al. [22–25]	Commercial-purity Cu and Al foil; dissimilar spot welds	Not stated; heat input independent of time, so assume constant coefficient of friction or shear stress	Surface friction and plastic deformation	Two cases: Johnson-Cook model with non-ultrasonic literature data; and modified J-C model calibrated to tensile tests, and compression tests with ultrasonic vibration	Temperature, sample deformation
Shen et al. [26]	Cu sheet; spot weld (battery tab-bus)	Temperature-dependent coefficient of friction	Constant heat input per unit area (calibrated using temperature data)	Cyclic plasticity combining isotropic and kinematic hardening, empirical ultrasonic softening, and temperature dependency	Temperature, weld zone geometry
Ngo et al. [27, 28]	Al; spot and seam welds	Assumed friction conditions not stated	Friction and bulk plasticity as functions of yield strength, time-dependent weld area, and welding parameters (downforce, amplitude, and frequency)	Yield stress linear in temperature; no strain-rate dependence	Temperature (accuracy sensitive to number of locations measured)

temperature-dependent, or pressure-dependent. In most cases, this value of friction coefficient is calibrated by various friction experiments within a range of temperature values to provide the heat input, which should be consistent with measured temperature. Throughout, the conditions in these experiments are assumed to be identical with the USW process.

Nonetheless, the direct measurement of transverse force (which is utilized to find out the coefficient of friction in the USW process) is complicated. The temperature- and clamping pressure- dependent friction models can be adopted for the dynamic simulation of interface temperature during the USW process. From the curve fitting, the relationship between the friction coefficient with temperature and pressure can be obtained. Furthermore, the vibration amplitude also influences the degree of frictional heating. It is applied with frictional stress to produce frictional heating. This vibration amplitude is difficult to measure experimentally, as the USW process yields very small displacement with high frequencies.

Investigators now comprehend that the frictional heating in thermo-mechanical modeling is directly dependent on the relative velocities of the contacting surfaces between the weld specimens. Meanwhile, horizontal velocities (along with the vibration direction) of the upper and lower weld specimens are generated by the motions of sonotrode and anvil tip (as USW is a dual-head wedge-reed system, both sonotrode and anvil oscillate in an anti-phase manner). Thus, the horizontal movements of the sonotrode and anvil tips are considered as inputs to the existing model, and they can be measured experimentally using laser vibrometers during the welding process. As there is always some noise during the experimental measurement process, the raw data of this wave cannot be used directly as input data for modeling. Instead, this experimental velocity ($V_i(t)$) data, along with frequencies and phases (if any), are utilized to find out the input displacements $x_i(t)$ of each tip concerning time (t). For this analysis, the frequency of this motion (ω_i) can be extracted from the experimentally measured velocity by using the fast Fourier transform technique. In this particular case, the ω_i is taken as 20.9 kHz. The time-dependent amplitude or the tip displacement can be expressed as:

$$x_i(t) = \frac{V_i(t)}{\omega_i} \mathrm{Sin}(\omega_i t - \phi_i) \qquad (6.105)$$

where ϕ_i is the phase angle. In the dual-head wedge-reed systems, the phase difference between sonotrode and anvil velocity is set as 90°. Finally, the resulting sonotrode and anvil displacements are considered as inputs to the thermo-mechanical modeling simulation. Figure 6.22 illustrates the displacement profiles obtained from laser vibrometers. According to the ultrasonic welding principle, the whole process is divided into three transitional stages (I, II, and III). In these three stages, the displacements of the two weld specimens as well as sonotrode tip are determined. Later, these displacements can be converted to velocity by dividing the weld time.

The later stage in the USW process involves fully coupled thermo-mechanical models that predict temperature and deformation responses simultaneously by providing the heat input through friction, plastic deformation, and elastic hysteresis. In this stage, the kinematic and frictional boundary conditions are applied at their respective locations. In USW, the hot plasticity constitutive model is particularly

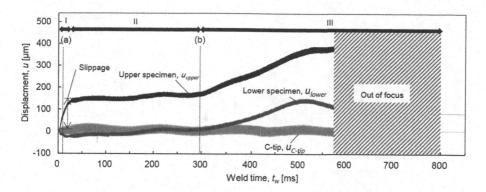

FIGURE 6.22 Comparison of measured and simulated velocity results of (a) upper weld sample, and (b) lower weld sample [29].

useful. It may be divided into two types: conventional and advanced models. Conventional models engage uniaxial hot flow stress models which can capture interface temperature and strain-rate-dependent flow stress with isotropic hardening. These types of models employ the Johnson-Cook equation. As the FE solution proceeds, the stresses are developed in the material, and it is plastically deformed. This plastic deformation develops heat internally in addition to the frictional heat previously discussed. Thus, this Johnson-Cook model is capable of providing an accurate solution, which can be expressed as

$$\sigma = [A + B(\varepsilon_p)^n](1 + C\ln\dot{\varepsilon}_0)(1 - T_0^m) \tag{6.106}$$

where σ is the stress in MPa, ε_p is the plastic strain, $\dot{\varepsilon}_0$ is the dimensionless plastic strain rate (1.0s^{-1}), and T_0 is the dimensionless temperature; A (MPa), B (MPa), C (0.002 for Al), n, and m are the constants. However, the complex USW process requires an advanced hot plasticity constitutive model which takes into account kinematic hardening due to cyclic deformation and acoustic softening (reduction in flow stress due to high-frequency ultrasonic vibration). Most of the modeling approaches in USW lack computational efficiency because they consider only 1,000 cycles due to high-frequency process. The computational efficiency of the problem can be improved in four different ways: (1) using explicit dynamics to predict the frictional heat flux, (2) treating this predicted heat flux as input to the implicit analysis of the whole USW process, (3) limiting the size of the model and mesh grids, (4) simplifying the 3D model to a 2D model. During the USW process, the material strength is reduced due to the acoustic softening effect, and a scaling factor is introduced to represent this reduction in the yield strength of the material ($\sigma_y(t)$) under the volumetric acoustic softening density (ρ_{US}). It is represented as

$$\sigma_y = \sigma_y(t) \times \exp(-n_2 \times \rho_{US}) \tag{6.107}$$

where n_2 is a material-related parameter (it is 0.155 mm^3/J for aluminum material).

The explicit module of ABAQUS® is applied to find out the interface temperature during USW of AA6111 aluminum alloy to DC04 steel/Ti6Al4V titanium alloy by

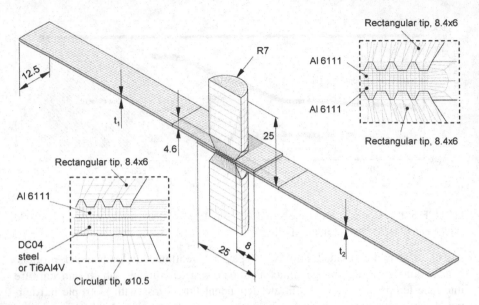

FIGURE 6.23   3D thermo-mechanical FE model [9].

employing a 3D thermo-mechanical model. The sheet thickness is considered accord-ing to the material combination used in the experiment. It utilizes 8-node brick (DC3D8) and 4-node tetrahedron (DC3D4) linear heat transfer elements for this thermo-mechanical analysis. Figure 6.23 shows the anvil and sonotrode tips with only half portions of the weld specimens. In this model, the lower specimen is placed on the anvil, and the upper specimen is in contact with the sonotrode knurls. The insets of this figure reveal the features of partial cross-sections of the weld region, exposing the geometric dimensions of the two types of sonotrodes (i.e., rectangular and circular). Meanwhile, the clamping pressure is applied on the sonotrode knurled tip, and the anvil tip is fixed in the vertical y-direction. The high-frequency ultrasonic vibration is applied on the sonotrode tip horizontally, and thus motion of the anvil tip also occurs. All the components in this model are considered as deformable 3D solid elements. The displacements of sonotrode and anvil knurls are selected as inputs to this model.

The motions of both upper and lower weld specimens are regulated by friction at the interface along with their contacts with the sonotrode and anvil tips. The coef-ficient of friction varies with the surface conditions as well as with the progres-sion of weld time. Thus, it is difficult to measure the coefficient of friction precisely with respect to weld time. In the above-described model, the coefficient of friction at all contiguous surfaces is taken as constant: i.e., 0.6 for the surfaces between sonotrode and anvil tips with the weld specimens, and 0.2 at the weld interface. The contact surfaces in this model are defined in LS-DYNA with a keyword CONTACT _AUTOMATIC _ SURFACE _ TO _ SURFACE _ THERMAL. In the developed FE model, the central region under the sonotrode knurl pattern is considered the major source of heat flux. The thermal properties such as specific heat and thermal conductivity are applied for the respective components of the model. The AA5754 aluminum alloy weld sample has 926 J/kg K specific heat and 205 W/mK thermal conductivity values. In comparison, the steel tips of the sonotrode and anvil have 480 J/kg K specific heat and 50 W/mK thermal conductivity values. The coefficient of thermal contact conductance of 2000 W/m²K is also applied. The convective boundary condition is applied on the weld specimen surfaces that are exposed to air. Usually, the coefficient of convective heat transfer is considered to be 30 W/m²K.

AZ31 magnesium alloy sheets are welded using an ultrasonic welding machine with dual sonotrodes of symmetric configuration. A 2D plane strain FE model is utilized for the thermo-mechanical modeling of this process, and it is carried out in two FE steps: (1) during the clamping period and (2) during sonotrode vibration. This explicit time integration analysis is followed in the simulation technique, as the process happens in a short time interval, and an implicit analysis of the whole process would involve far more computational time and cost. As discussed earlier, the heat generation during USW relies on the frictional heating and plastic deformation at different contact surfaces. Thus, the coefficient of friction for the frictional heating is considered in this study as a constant value, in addition to the values obtained from the polynomial relationships between friction coefficients with temperature and pressure values. The variable coefficient of friction values are implemented in this model by a user subroutine VFRIC or in a tabular form in the ABAQUS® interaction property module. The forced displacements of two sonotrodes are considered the input loading condition, and this displacement condition can be experimentally quantified by laser vibrometers or digital image correlation (DIC) with a high-speed camera set-up. This numerical analysis employs ABAQUS® 3D FE software. The 2D model is composed of CPE4RT four-node quadrilateral elements with reduced integration and hourglass control. Figure 6.24 illustrates this 2D model. The tool steel

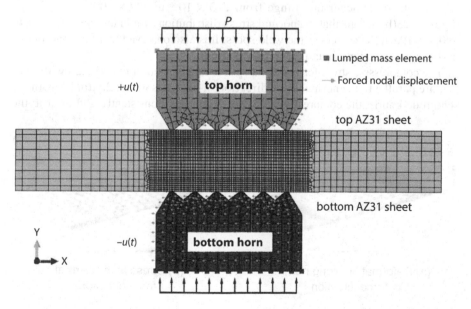

FIGURE 6.24 A 2D FE model with meshing for thermo-mechanical simulation of the USW process [30].

horn is regarded as an elastic body due to its superior strength, higher than that of the weld specimens. Apart from the sonotrode displacement and coefficient of friction, the amplitude of weld specimens, heat flux, thermal conductivity, specific heat, and convective heat transfer coefficient are defined for all the interfaces. The dynamic effects of the sonotrode can be captured fully by adding a lumped mass on the top surface of the modeled tip to save computational time and cost.

During the simulation of this process, the first step is assumed to be 10 msec. In this time, the load is ramped up from zero to predefined pressure P MPa. In the second step, the displacement is applied to the sonotrode per the previously described method, and the vibration amplitudes of the upper and lower weld specimens are also applied per the following equations:

$$x_{upper}(t) = x_0 \mathrm{Sin}(2\pi f t) \qquad (6.108)$$

$$x_{lower}(t) = x_0 \mathrm{Sin}(2\pi f t + \phi) \qquad (6.109)$$

where x_0 is the experimentally acquired vibration amplitude, f is vibration frequency, t is the weld time, and ϕ is the phase angle between the two sonotrodes.

The plastic deformation is quantified using the von Mises yield criterion with the linear kinematic hardening rule.

6.3 VALIDATION OF SIMULATED RESULTS

The deformation and normal stress distribution on the weld samples along the normal (Y-axis) direction are shown in Figure 6.25. It may be observed from Figure 6.25(a) that due to the penetration of the sonotrode knurls, the deformations at the weld spots occur in a range from 253×10^{-6} to 327×10^{-6} m. Likewise, Figure 6.25(b) reveals that the normal stress distributions are in the range from 0.46 to 0.58 MPa at the centers of the weld spots. It is also observed that the non-contact portions withstand minimum compressive stress.

The compressive stress for one weld spot is represented in Figure 6.26 with lines that are parallel to the normal force direction. It indicates up to 220 μm beneath the sonotrode knurls; the compressive stress is decreasing consistently. But after it, the

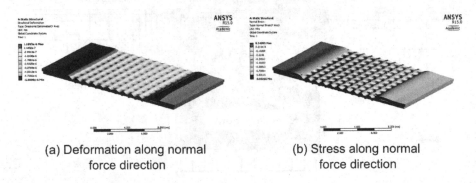

(a) Deformation along normal   (b) Stress along normal
force direction   force direction

FIGURE 6.25 Deformation and stress distributions of Al-Cu weld coupons along the normal direction (Y-axis) during the delay time.

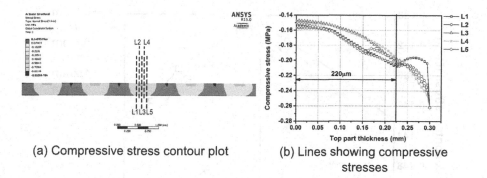

(a) Compressive stress contour plot

(b) Lines showing compressive stresses

FIGURE 6.26 Compressive stress distributions along lines parallel to normal force direction for one weld spot.

changes are erratic. This means that with up to 220 μm thickness of the top part, the compressive force has some effect.

Figure 6.27 presents a comparison graph of the total compressive load obtained by FE analysis and lathe tool dynamometer. The FE analysis results show a nature of curves similar to what was produced by the experiments. Thus, it can be said that the present finite element model is capable of predicting the compressive stress distribution with good accuracy.

The effect of top-part thickness on the contact stresses between the sheets can be explored through this FE analysis. The importance of this model lies in determining the maximum thickness of the sheets that can be welded using current ultrasonic welding machines. Thus, a nominal clamping pressure of 0.2 MPa is applied during the analysis. In this case, the rectangular knurled sonotrode tip is pressed to the Al-Cu sheets, which in turn are fixed on a circular flat anvil. The area of the anvil is

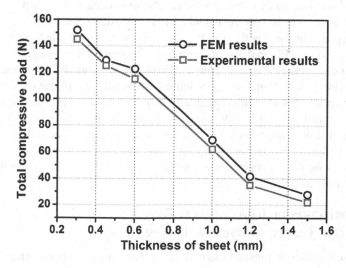

FIGURE 6.27 Comparison of total compressive load obtained by FE analysis and lathe tool dynamometer.

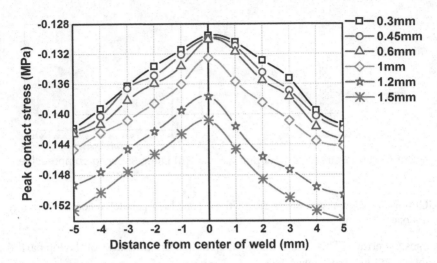

FIGURE 6.28 Distribution of peak contact stress on the bottom sheet for various thicknesses of the top part.

larger than the area of the weld tip. The peak compressive stress distributions on the bottom sheet (Cu) are illustrated graphically (Figure 6.28).

It can be clearly observed that the maximum contact stress is significantly decreased by the increase of sheet thickness. It is also revealed that for top-sheet (Al) thickness up to 0.6 mm, the bottom worksheet experiences almost uniform compressive stress in the weld zone. But when the thickness is increased further, above 1 mm thickness of the workpiece, the peak contact stress is drastically reduced. In real welding conditions, it is not possible to achieve welds with sheets thicker than 1 mm at these machine settings.

During the welding process, two forces are acting simultaneously: normal force and shear force. Due to these two forces, deformation occurs in the top sheet as it is in direct contact with the sonotrode. Only a negligible deformation is observed in the bottom sheet as the sheet is fixed to the anvil. Thus, to find the deformation during the welding process, FE analysis has been done for four different surface conditions. The resulting deformation contour plots are displayed in Figure 6.29. One can observe from the weld cross-sectional images (Figure 6.29(b), (d), (f), (h)) that the deformation at the weld spots (dark blue color) gradually increases with the coefficient of friction. Even with a marginal change in deformation, the results are still valuable because the plastic deformation and generation of heat are dependent on it. These deformations at the weld interface are the reasons for under, good, and over weld conditions. Likewise, the distortions on the top sheet formed by the sonotrode knurls are responsible for tip sticking and extrusion phenomena.

6.3.1 PREDICTION OF TEMPERATURES AT DIFFERENT ZONES DURING THE WELDING PROCESS

As discussed earlier, a transient thermal simulation study has been carried out, and the prediction results are displayed. Figures 6.30 to 6.33 display the contour plots and isolines of temperature for various surface conditions: as lubricating, normally polished, electrolytically polished, and emery polished, respectively. In each figure,

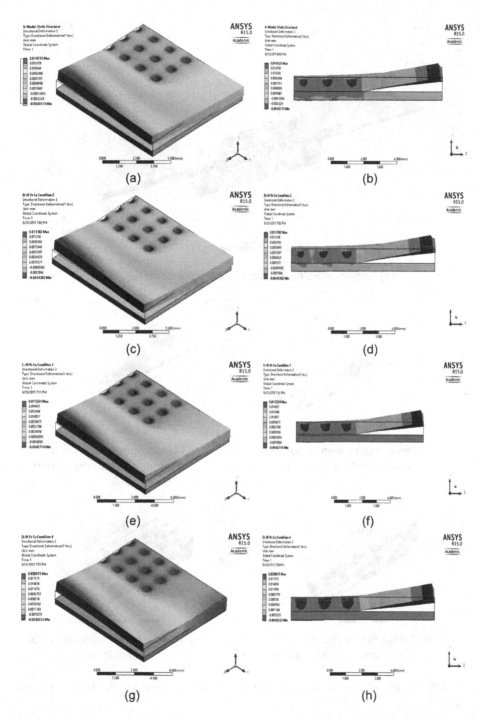

FIGURE 6.29 Peak contact stress prediction on a quarter-section of weld samples (a) for lubricating condition; (c) for normally polished condition; (e) for electrolytically polished condition; (g) for emery polished condition; (b), (d), (f), and (h) are cross-sectional images of (a), (c), (e), and (g) showing distribution of contact stress.

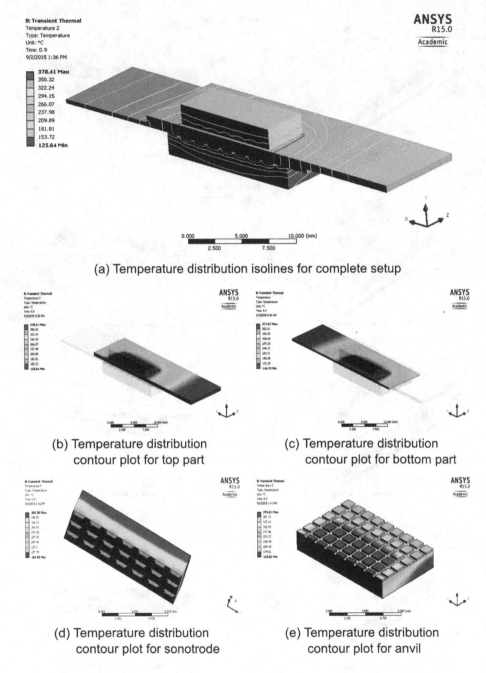

(a) Temperature distribution isolines for complete setup

(b) Temperature distribution
contour plot for top part

(c) Temperature distribution
contour plot for bottom part

(d) Temperature distribution
contour plot for sonotrode

(e) Temperature distribution
contour plot for anvil

FIGURE 6.30 Contour plots and isolines of temperature for lubricating surface condition.

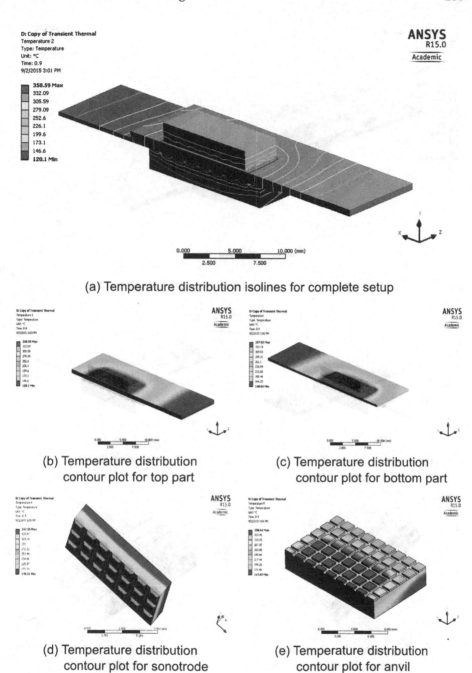

(a) Temperature distribution isolines for complete setup

(b) Temperature distribution contour plot for top part

(c) Temperature distribution contour plot for bottom part

(d) Temperature distribution contour plot for sonotrode

(e) Temperature distribution contour plot for anvil

FIGURE 6.31 Contour plots and isolines of temperature for normally polished surface condition [1].

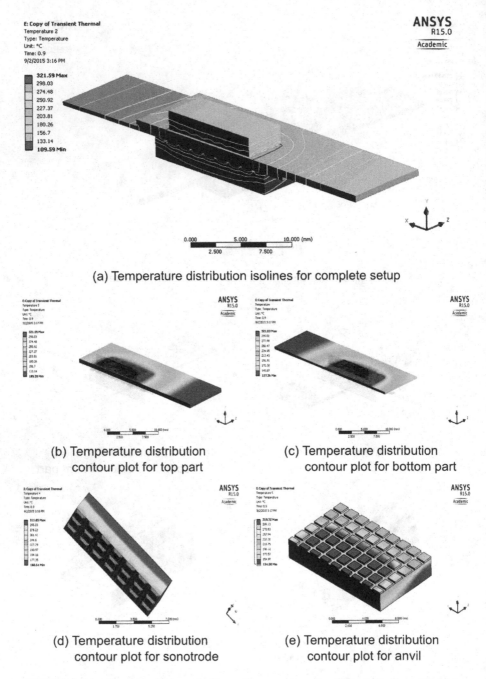

(a) Temperature distribution isolines for complete setup

(b) Temperature distribution
contour plot for top part

(c) Temperature distribution
contour plot for bottom part

(d) Temperature distribution
contour plot for sonotrode

(e) Temperature distribution
contour plot for anvil

FIGURE 6.32   Contour plots and isolines of temperature for electrolytically polished surface condition.

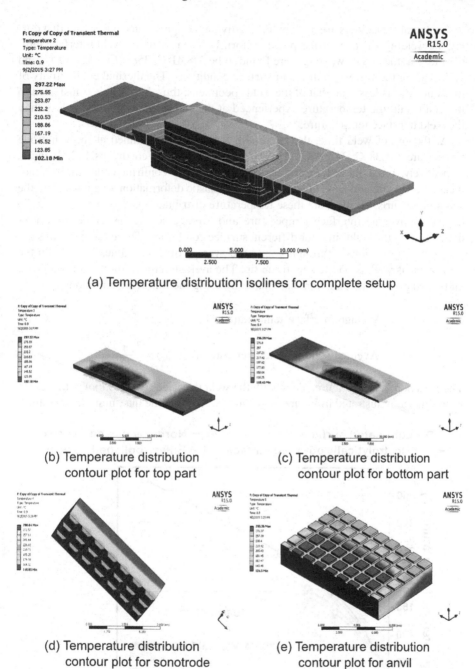

(a) Temperature distribution isolines for complete setup

(b) Temperature distribution contour plot for top part

(c) Temperature distribution contour plot for bottom part

(d) Temperature distribution contour plot for sonotrode

(e) Temperature distribution contour plot for anvil

FIGURE 6.33 Contour plots and isolines of temperature for emery polished surface condition.

the temperature isolines are distributed evenly on the specimens, indicating that during the welding, the heat is dissipated uniformly. The maximum weld interface temperatures at the end of welding were found to be 378.41°C, 358.59°C, 321.59°C, and 297.22°C for the corresponding four surface conditions. The thermal conductivity of the sonotrode is less than that of the weld specimens; thus, less heat is transferred to it. As a result, the temperature experienced at the end of the sonotrode is less than the weld interface temperature.

At the end of weld time, the maximum temperatures obtained at the sonotrode tip face are 366.38°C, 347.35°C, 311.85°C, and 288.04°C for conditions 1, 2, 3, and 4, respectively. From these temperature predictions, it is confirmed that the thermo-mechanically affected zone (TMAZ) is on the plastic deformation area just below the sonotrode knurls. To illustrate these temperature distributions, Figures 6.34 and 6.35 depict the average interface temperature and average sonotrode temperature predicted during the weld time for different surface conditions. These temperatures are predicted on positions 1 through 4, and the average of those values is taken in the weld zone as well as on the sonotrode tip. The average temperatures for positions 1 and 2 and positions 3 and 4 are calculated by using the following equations:

$$\text{Average interface temperature } (T_{\text{weld interface}}) = \frac{T_1 + T_2}{2} \qquad (6.110)$$

$$\text{Average sonotrode temperature } (T_{\text{sonotrode}}) = \frac{T_3 + T_4}{2} \qquad (6.111)$$

The variation of temperature, along with the weld interface (X-direction) at the end of weld time, are illustrated in Figure 6.36. One can see that the maximum temperatures

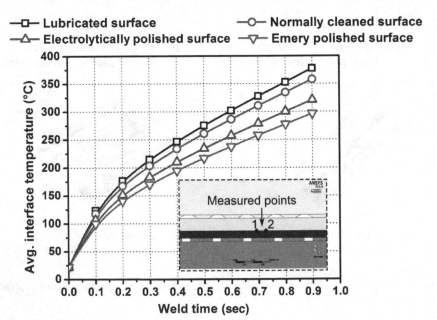

FIGURE 6.34 Predicted average weld interface temperatures for various surface conditions.

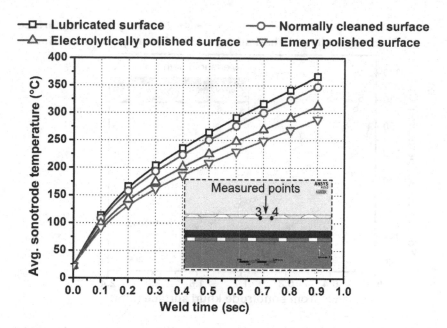

FIGURE 6.35   Predicted average sonotrode temperatures for various surface conditions.

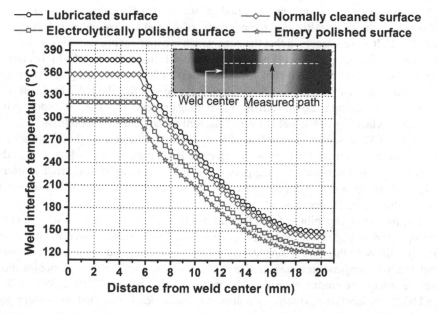

FIGURE 6.36   Variation of temperature along the weld interface (X-direction) at the end of weld time for various surface conditions.

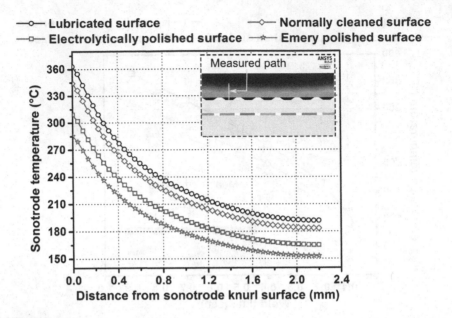

FIGURE 6.37 Variation of temperature in sonotrode (Y-direction) at the end of weld time for various surface condition.

for each surface conditions are relatively constant up to 5.7 mm from the center of the weld. Thus, this portion signifies that it is under the sonotrode knurl; that is, in the TMAZ where the area is influenced by both plastic deformation and acoustic softening. Meanwhile, the recrystallization temperatures of Al and Cu are measured as 240°C and 370°C. It is well accepted that the zone which is experiencing heat higher than the recrystallization temperature is known as the heat-affected zone (HAZ). In this analysis, after the TMAZ begins to cool, the temperature falls consistently up to 20 mm and reaches a lowest temperature of 120°C. Thus, it can be said that the HAZ for Cu as well as Al is developed up to 20 mm during the welding process. This HAZ depends on the thermal conductivity of the weld materials as well as their surface conditions.

Figure 6.37 presents the variation of temperature in the sonotrode (Y- direction) at the end of weld time. One can notice that the HAZ present in it is less than in the workpiece material, because of the poor thermal conductivity of sonotrode material (D2 steel) compared to work material. On moving in the Y-direction, the temperature is also gradually decreased.

The predictive capability of this thermo-mechanical model can be tested by comparing the prediction results with experimental ones. Figure 6.38 shows the comparison graph of the weld interface temperature for various surface conditions. The maximum experimental temperatures measured at the end of weld time by thermocouples show a close fit with the predicted results. The errors are obtained as 0.80%, 2.35%, 0.78%, and 0.72% for lubricating, normally polished, electrolytically polished, and emery polished surface conditions, respectively. Thus, it is confirmed that the boundary conditions that were applied to this thermo-mechanical problem are reasonable.

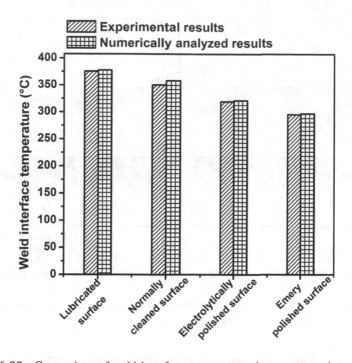

FIGURE 6.38 Comparison of weld interface temperatures between experimental results and numerical results for different surface conditions.

Figure 6.39(a) shows the simulated temperature distribution in the weld specimen, as well as in the tools, in a 3D view during the USW process. Figure 6.39(b) shows the quarter portion of the model to reveal the temperature profile at the weld spot. It can be inferred from this figure that the temperature is highest at the center of the weld spot (red color), and that it decreases gradually in all possible directions away from the weld spot. However, the temperature of the sonotrode/anvil is less than that of the specimens due to their low thermal conductivity properties.

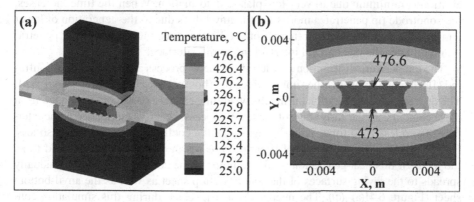

FIGURE 6.39 (a) 3D plot of temperature distribution, (b) quarter section showing temperature contours at the end of weld time (red color represents central weld zone) [31].

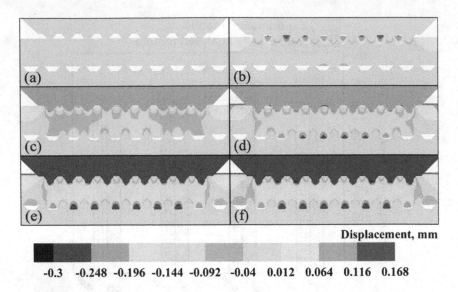

Displacement, mm

-0.3   -0.248   -0.196   -0.144   -0.092   -0.04   0.012   0.064   0.116   0.168

FIGURE 6.40   Plastic strain and deformation of weld samples under sonotrode and anvil indentions at various weld times [31].

Figure 6.40 displays the plastic strain simulation results, which consist of deformation shape of the weld samples and the level of plastic strain at the end of weld time. The red color in this model represents the zones where the plastic strain is the highest. Moreover, the distribution of these regions can be noticed just below the sonotrode tip; they eventually decrease with each successive row. It can be inferred from this figure that the rounded plastic strain contours are generated from the regions where the knurls of the sonotrode and anvil are indented on the sheets. However, the physics behind the formation of these types of patterns are still unknown. It is also quite interesting to observe that the rate of penetration of these sonotrode knurls into the weld specimen increases with a rise in weld time. At lower weld time, the penetration is minimum due to very low plastic deformation. When the time increases, the sonotrode tip penetrates more than the anvil does due to the generation of lower temperature and ultrasonic softening of the bottom specimen. The deeper penetration facilitates the mechanical interlocking at the interface.

The temperature distribution profiles of the USW process are captured by an infrared (IR) camera (top), and the simulated results (bottom) are compared in Figure 6.41. The degree of heat generation during the weld time can be clearly comprehended from these contours. At the early stage of the USW process, heat is accumulated at the interface due to less weld time. Thus, in this stage, the coefficient of friction is also less. However, when this coefficient of friction increases in due course of the weld time, the locus of this heat-generation profile remains at the central zone, and it gradually spreads to the faying surfaces of the sonotrode-top sheet as well as the anvil-bottom sheet (Figure 6.41(a)–(c)). The interesting fact revealed during this simulation concerns the constant coefficient of friction used in the cases of Figure 6.41(a)–(d): the friction coefficient of Figure 6.41(c) severely underpredicts the interface temperature.

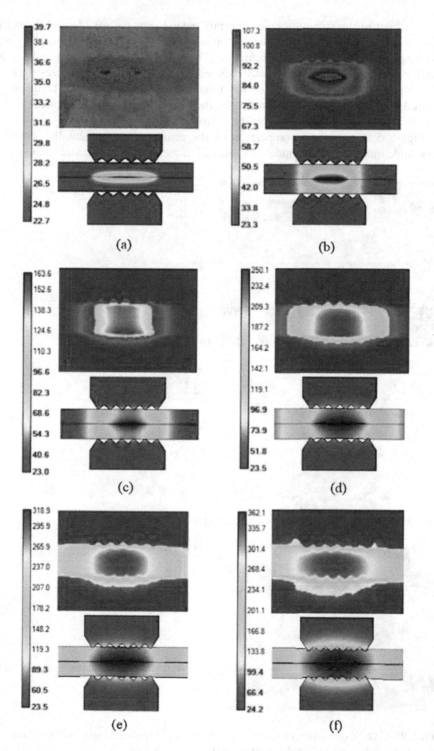

FIGURE 6.41 Comparison of IR camera-measured interface temperature distribution with simulated results at different coefficient of friction values [30].

Thus, a higher-value friction coefficient is utilized in Figure 6.41(d). Meanwhile, as the sonotrode and anvil knurls indent on the top and bottom sheets, a large amount of heat is also produced due to plastic deformation. The faying surface of the material is softened, and the yield strength is greatly reduced. Figure 6.41(e) and (f) utilizes the data from the polynomial relationship between the coefficient of friction and temperature as well as pressure. In these two instances, the temperature profile is elliptical, and it accurately predicts the actual temperature data.

Figure 6.42 presents a comparison of the results of experimentally observed sonotrode indention with simulated values from the thermo-mechanical analysis. It

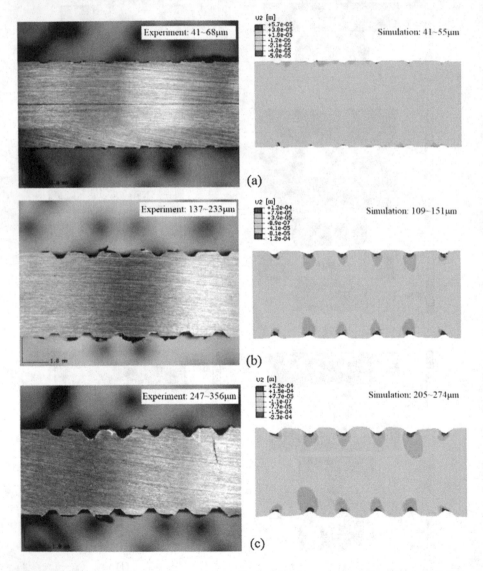

FIGURE 6.42 Comparison of experimentally measured sonotrode and anvil knurls indention depth with simulated results at different weld energies [30].

is obvious in Figure 6.42(a) that there is no bonding at the interface, due to low weld energy values. However, a good metallurgical bond can be perceived in the other two conditions (Figure 6.42(b) and (c)) as the weld energy is raised. Meanwhile, the increase in weld energy values results in a rise in interface temperature, and the material is softened. Both the experimental and simulated results exhibit the corresponding increment in the sonotrode and anvil knurls indention magnitude with the increase in weld energy.

6.4 EFFECT OF ACOUSTIC SOFTENING PARAMETER

As an illustration, two materials are chosen: AA1100 aluminum sheets and UNS C10100 copper sheets. Of these sheets, aluminum has the lower yield strength. Thus, it is expected that much plastic deformation will occur in the aluminum sheet, and the finite element modeling process is simplified by acknowledging this. For numerical analysis, a 3D finite element model was analyzed by the explicit dynamics solver of ANSYS® 15.0. Figure 6.43 depicts the boundary conditions of the numerical model used for acoustic softening analysis. In this model, the types of mesh elements for sonotrode, anvil, and work materials were selected exactly as previously described for the improvement of computational efficiency. The boundary conditions for surfaces of the top and bottom work materials were defined using the penalty contact method. The contact between the sonotrode and the top part was determined by the penalty friction method with a coefficient of friction of 1. This indicates that the relative motion between the top sheet and sonotrode is restricted. A similar penalty friction method was applied to the bottom sheet and anvil. The coefficient of friction between the two sheets was applied according to the surface conditions. A zero X- and Y-displacement was assigned to one of the edges of the top sheet to

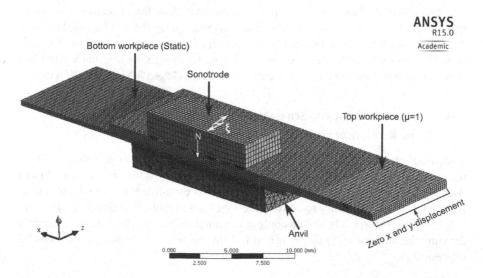

FIGURE 6.43 Boundary conditions of the numerical model used for acoustic softening analysis [1].

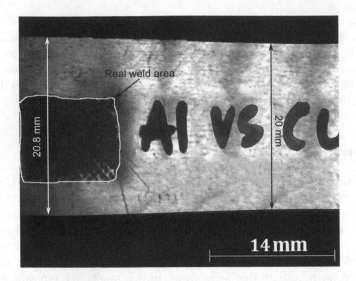

FIGURE 6.44 Sample image used to calculate the average width after welding.

constrain the displacement of the sheet in that direction. To increase computational efficiency, this thermo-mechanical model was executed only for 0.001 sec. This time was selected after simulating the model and observing the time at which adequate plastic deformation took place.

6.4.1 MEASUREMENT OF SHEET WIDTH INCREMENT

An optical image of each welded sample produced with different surface conditions was selected to measure the average width increase near the weld spot. This was done by using Image J software. Figure 6.44 displays a sample image used to calculate the average width after welding. Notice that although the original width ($\mathcal{W}_o$) of the specimen was 20 mm, after welding the width ($\mathcal{W}_o+\Delta\mathcal{W}$) became 20.8 mm.

6.4.2 EFFECT OF ACOUSTIC SOFTENING PARAMETERS ON THE REDUCTION OF YIELD STRENGTH

The effect of an acoustic softening parameter using Eq. (6.84) was described by using the thermo-mechanical model. To validate it, steady-state temperature (T) and $\Delta\mathcal{W}/\mathcal{W}_o$ were calculated for the Al-Cu sheets. To establish the acoustic softening parameter, it was necessary to convert the vibration amplitude and weld pressure to a dimensionless form. The dimensionless vibration amplitude (Ψ) can be derived by dividing the amplitude (ξ) by the initial foil thickness (t). Mathematically, it can be expressed as:

$$\psi = \frac{\xi}{t} = \frac{68\times10^{-3}}{0.7} = 0.09 \qquad (6.112)$$

Likewise, the dimensionless weld pressure (P) can be written as the weld pressure (WP) divided by the yield strength (σ_y) of the material (σ_y = 34 MPa at room temperature). Therefore, the mathematical expression is:

$$P = \frac{WP}{\sigma_y} = \frac{0.38}{34} = 0.01 \qquad (6.113)$$

The coefficient of friction values described previously are used in this section, and the acoustic softening parameter (β) was solved by decreasing the β value incrementally. This was carried out until the predicted $\Delta \mathcal{W}/\mathcal{W}_0$ and experimental $\Delta \mathcal{W}/\mathcal{W}_0$ were a good match. To shed light on the effect of acoustic softening, Figure 6.45 represents the experimental as well as predicted $\Delta \mathcal{W}/\mathcal{W}_0$ with three β values for different surface conditions.

The various β values were fed to the explicit dynamics model of ANSYS® and the corresponding deformation in the width of the sheet was observed. For all the surface conditions, Ψ is 0.09 and P is 0.01. Thus, the model using $\beta = 1$ severely underpredicts the sheet width increase as comparable to $\beta = 0.13$. As previously described, when $\beta = 1$, no acoustic softening happens. However, if this value is decreased, the acoustic softening will play a vital role. It clearly perceives the figure that the majority of the material softening and reduction in yield strength happened due to acoustic softening effect. To test the model results for different surface conditions, it is essential to compare the modeled decrease in the thickness of the top sheet and the experimentally measured sheet thickness. This checking has been done after fixing

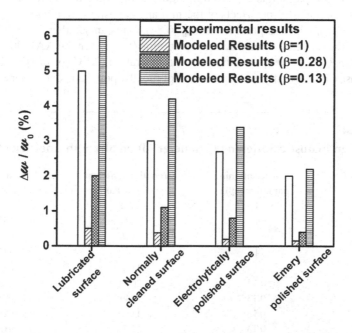

FIGURE 6.45 Comparison of $\Delta \mathcal{W}/\mathcal{W}_0$ between modeled and experimental results using three acoustic softening parameter (β) values for various surface conditions.

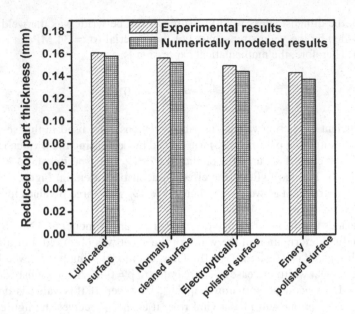

FIGURE 6.46 Comparison of reduced top-part thickness values between modeled and experimentally measured results for various surface conditions.

the β value for minimization of error between the modeled and experimental $\Delta w / w_o$ results. This comparison is displayed in Figure 6.46 for 0.7Al-0.4Cu sheets.

Table 6.4 compares the effects of thermal and acoustic softening on the strength coefficient (K_s), which is provided in Eq. (6.84). One can see that in the lubricating condition, there is a significant acoustic softening (53.73%) in the AA1100 aluminum sheet at 375.37°C as compared to its room temperature (20°C) properties. Likewise, in the cases of normally polished, electrolytically polished, and emery polished

---

TABLE 6.4

Effect of an Acoustic Softening Parameter (β) on Strength Coefficient (K_s)

Temperature (T_s) (°C)	Acoustic softening parameter (β)	Strength coefficient (K_s) (MPa)	% K_s reduction from room condition
20	1	157.8	0
375.37	1	73	53.73
350.35	1	85.6	45.75
319.09	1	98.2	37.76
295.08	1	110.8	29.78
20	0.13	20.51	87
375.37	0.13	9.49	93.98
350.35	0.13	11.12	92.95
319.09	0.13	12.76	91.91
295.08	0.13	14.40	90.87

surface conditions, acoustic softening effects of 45.75%, 37.76%, and 29.78% were observed. When the β = 0.13, a considerable reduction in the yield strength of the material for different surface condition values—87%, 93.98%, 92.95%, 91.91%, and 90.87%—was found while doing a comparison study with the normal temperature properties.

These experiments show that coupled thermo-mechanical modeling can predict the temperature and deformation of weld materials. Thus, a welder can use these modeling results to optimize the process parameters and get the desired joint strength with good weld quality. Moreover, this model can predict the results at extreme weld conditions that cannot be achieved by an experiment. Therefore, it not only provides insight into the safety of the welding process, but also offers an opportunity to increase the capacity and efficiency of the USW machine. From the deformation simulation results, the indention depth of the sonotrode knurls can be understood. This information will help to determine the maximum depth of penetration on the sheet by the sonotrode knurls during an experiment. The commonly used trial-and-error method may increase the chance of sonotrode tip damage, which will cause a significant delay in the welding process and incur high repair costs.

REFERENCES

1. Satpathy MP, Mohapatra KD, Sahoo SK. Ultrasonic spot welding of Al-Cu dissimilar metals: a study on parametric influence and thermo-mechanical simulation. Int J Model Simul 2018;38:83–95. doi:10.1080/02286203.2017.1395198.
2. Langenecker B. Effects of ultrasound on deformation characteristics of metals. Sonics Ultrason IEEE Trans 1966;13:1–8.
3. Hockett JE. Relating the flow stress of aluminum to strain, strain rate and temperature. AIME MET SOC Trans 1967;239:969–76.
4. Siddiq A, El Sayed T. Acoustic softening in metals during ultrasonic assisted deformation via CP-FEM. Mater Lett 2011;65:356–9.
5. Huang H, Chang BH, Du D. Effect of superimposed ultrasound on mechanical properties of copper. Mater Sci Technol 2011;27:1117–22.
6. Hung J-C, Hung C. The influence of ultrasonic-vibration on hot upsetting of aluminum alloy. Ultrasonics 2005;43:692–8.
7. Siddiq A, El Sayed T. A thermomechanical crystal plasticity constitutive model for ultrasonic consolidation. Comput Mater Sci 2012;51:241–51.
8. Siddiq A, El Sayed T. Ultrasonic-assisted manufacturing processes: variational model and numerical simulations. Ultrasonics 2012;52:521–9.
9. Jedrasiak P, Shercliff HR. Finite element analysis of heat generation in dissimilar alloy ultrasonic welding. Mater Des 2018;158:184–97. doi:10.1016/j.matdes.2018.07.041.
10. Elangovan S, Semeer S, Prakasan K. Temperature and stress distribution in ultrasonic metal welding—An FEA-based study. J Mater Process Technol 2009;209:1143–50.
11. Sooriyamoorthy E, Henry SPJ, Kalakkath P. Experimental studies on optimization of process parameters and finite element analysis of temperature and stress distribution on joining of Al–Al and Al–Al$_2$O$_3$ using ultrasonic welding. Int J Adv Manuf Technol 2011;55:631–40.
12. Siddiq A, Ghassemieh E. Thermomechanical analyses of ultrasonic welding process using thermal and acoustic softening effects. Mech Mater 2008;40:982–1000.
13. Siddiq A, Ghassemieh E. Theoretical and FE analysis of ultrasonic welding of aluminum alloy 3003. J Manuf Sci Eng 2009;131:41007.

14. Kim W, Argento A, Grima A, Scholl D, Ward S. Thermo-mechanical analysis of frictional heating in ultrasonic spot welding of aluminium plates. Proc Inst Mech Eng Part B J Eng Manuf 2011;225:1093–103.
15. Doumanidis C, Gao Y. Mechanical modeling of ultrasonic welding. Weld J 2004;83:140–6.
16. Gao Y, Doumanidis C. Mechanical analysis of ultrasonic bonding for rapid prototyping. J Manuf Sci Eng 2002;124:426–34.
17. Zhang CS, Li L. Effect of substrate dimensions on dynamics of ultrasonic consolidation. Ultrasonics 2010;50:811–23.
18. Zhang C, Li L. A friction-based finite element analysis of ultrasonic consolidation. Weld J 2008;87:187.
19. Zhang C (Sam), Li L. A coupled thermal-mechanical analysis of ultrasonic bonding mechanism. Metall Mater Trans B 2009;40:196–207. doi:10.1007/s11663-008-9224-9.
20. Lee D, Kannatey-Asibu E, Cai W. Ultrasonic welding simulations for multiple layers of lithium-ion battery tabs. J Manuf Sci Eng 2013;135:61011.
21. Lee D, Cai W. The effect of horn knurl geometry on battery tab ultrasonic welding quality: 2D finite element simulations. J Manuf Process 2017;28:428–41.
22. Chen KK, Zhang YS. Numerical analysis of temperature distribution during ultrasonic welding process for dissimilar automotive alloys. Sci Technol Weld Join 2015;20:522–31.
23. Chen KK, Zhang YS, Wang HZ. Study of plastic deformation and interface friction process for ultrasonic welding. Sci Technol Weld Join 2017;22:208–16.
24. Chen K, Zhang Y. Mechanical analysis of ultrasonic welding considering knurl pattern of sonotrode tip. Mater Des 2015;87:393–404.
25. Chen K, Zhang Y, Wang H. Effect of acoustic softening on the thermal-mechanical process of ultrasonic welding. Ultrasonics 2017;75:9–21.
26. Shen N, Samanta A, Ding H, Cai WW. Simulating microstructure evolution of battery tabs during ultrasonic welding. J Manuf Process 2016;23:306–14.
27. Ngo T-T, Huang J-H, Wang C-C. The BFGS method for estimating the interface temperature and convection coefficient in ultrasonic welding. Int Commun Heat Mass Transf 2015;69:66–75.
28. Huang J-H, Ngo T-T, Wang C-C. HSDM and BFGS method for determining the heat generation and range of heat distribution in 2-D ultrasonic seam welding problems. Numer Heat Transf Part B Fundam 2016;69:48–67.
29. Sasaki T, Sakata Y, Watanabe T Effect of tool geometry on ultrasonic welding process. IOP Conf Ser Mater Sci Eng 2014;61. doi:10.1088/1757-899X/61/1/012006.
30. Huang H, Chen J, Lim YC, Hu X, Cheng J, Feng Z, et al. Heat generation and deformation in ultrasonic welding of magnesium alloy AZ31. J Mater Process Technol 2019;272:125–36. doi:10.1016/j.jmatprotec.2019.05.016.
31. Li H, Cao B, Liu J, Yang J. Modeling of high-power ultrasonic welding of Cu/Al joint. Int J Adv Manuf Technol 2018;97:833–44.

7 Future Research Trends in Ultrasonic Spot Welding of Dissimilar Metal Sheets

Ultrasonic spot welding (USW) has gained an abundance of consideration for its ability to join dissimilar metal and alloys with widely different chemical, physical, and mechanical properties, which cannot be achieved through any conventional fusion welding process. The implementation of USW technique in automotive industries has proven a promising method to deliver defect-free joints in Li-ion battery manufacturing and overall weight reduction of the automobile. However, the industrial applications of USW are still limited, and several challenges must be met to achieve widespread availability of the excellent, robust, and reliable welds possible with this system.

7.1 CONTROLLING FORMATION OF UNDESIRED INTERMETALLIC COMPOUNDS

The formation of intermetallic compounds (IMCs) has been observed during the USW of most metal combinations, such as Al-Cu, Al-Mg, Al-steel, etc. The rate of formation and growth of these IMCs depends upon the interatomic diffusion rate during the welding process. Presently there are no mechanisms available to prevent the formation of IMCs. However, the thickness of IMCs can be reduced with a reduction in the heat input due to friction and plastic deformation. It has been shown in the literature that weld strength can be improved marginally with a decrease in IMC thickness. However, the formation of strong and less brittle IMCs can be attained by alloying these base metals or the introduction of an interlayer, which ultimately leads to increases in weld strength. Thus, attention should be devoted in the future to understanding IMC formation and its growth mechanisms, and exploring the addition of a third-party metal layer into alloying procedures that can retard the diffusion rate and produce less detrimental IMCs. Moreover, the use of adhesives between the sheets may also reduce the typical interatomic diffusion process, so this should be researched further. IMCs can also be deteriorated by controlling the bond mechanisms and plastic flow of materials during the welding process, which relies upon the design of the anvil and horn geometries.

7.2   ENHANCING WELD STRENGTH IN THICKER SHEETS

Most ultrasonic spot welders work at 20 kHz of ultrasonic frequency, with 2.5- to 3.5-kW power range. Such systems are capable of producing satisfactory results in joining of thinner sheets, but face difficulties when used on advanced materials and thicker sheets. Thus, the further development of existing systems is necessary to tackle these issues, and it can be achieved through the creation of more powerful transducers. In some of the automotive industries, USW systems of 5 kW–6 kW ultrasonic power operating at 20 kHz are working, but they are specifically designed for a definite purpose. If this trend is to increase, then such systems may reach a higher level of 10 kW. Another approach to increase the power of USW is to couple two or more transducers to the weld reed, which will provide better results due to the synchronous effect of the individual transducers. An example of such a system is the ultrasonic torsional welder. Furthermore, ultrasonic vibration can be applied to the horn as well as the anvil by the two independent transducers; this setup can be regarded as an alternative to achieve satisfactory welds. However, the design of these systems is quite complicated due to the phase and anti-phase effect.

7.3   CORROSION BEHAVIOR STUDY OF WELD JOINTS

The cracking of structural components due to corrosion, galvanic corrosion, and residual stress corrosion is a potential issue in the automobile sector, because such cracking compromises the service life of the vehicle. It puts another challenge in front of the welder to join dissimilar metals for various industrial applications, although the parts that are exposed to the environment are shielded against corrosion by painting or coating. Therefore, it is crucial to analyze the performance of ultrasonically welded components when these parts are subjected to galvanic corrosion. It is well known that minimal residual stress is produced during the USW process as compared to the conventional fusion welding process. The residual stress analysis is more complicated when the two materials have vastly different melting points and thermal conductivities. As some of the parts of the automobile undergo thermo-mechanical fatigue stress during the automobile's service life, it is essential to investigate the cracking behavior of welds due to corrosion.

7.4   CONTROL OF TOOL WEAR

Wear of the horn and anvil becomes critical when the USW process is taking place between high-melting-point and low-thermal-conductivity materials. Although it has been shown that a sound weld can be achieved without much penetration of the knurl tips into the nonferrous metals, it is usually accepted that shearing and dispersion of the asperities take place when the plunging of the knurls happens on the ferrous materials. Otherwise, weak metallurgical bonding, mechanical interlocking, and thicker IMC layer may occur at the weld interface. Thus, the analysis of tool wear after prolonged use with ferrous materials should be performed, so the effects of coating on the tool tips can be revealed.

7.5 JOINING OF HARDER MATERIALS

An adequate amount of plastic deformation with proper interdiffusion of metals is needed to create the good mechanical performance of joints. These microstructural properties are dependent upon various material properties, asymmetric distribution of temperature profile, and the stacking orientation of the materials. It turns out to be even more challenging when one tries to join a softer nonferrous metal and a harder ferrous metal by USW. Some previous literature has reported on efforts that have been made to weld these types of materials by employing the hybridization of welding techniques. These hybrid welding techniques are often incorporated with heat sources like arc, electric resistance, or inductance to soften the hard ferrous materials so that the rate of interdiffusion can be increased. More effort is required to explore thoroughly the capacity of this approach as an industrial technique.

7.6 PROSPECTIVE OPPORTUNITIES IN AEROSPACE, MEDICAL, ENERGY, AND AUTOMOTIVE SECTORS

Though there are many challenges that must be dealt with before the USW technique can be implemented commercially in industries, the welding of previously nonweldable materials made possible by the invention of the USW technique gives insight and opportunity to these industries. Aerospace, medical, electronics, automotive, and marine sectors need high-strength, lightweight, and conductive materials. These industries can all benefit from this dissimilar USW technique. USW has already been implemented in the automobile industries for mass production of dashboards, Li-ion batteries, and structural components. As the lesser electrical resistance of ultrasonically welded joints is extremely preferred for a good-quality battery assembly, more attention in this direction is required. Dissimilar welding of steels, magnesium alloys, and nickel-based superalloys may open up new opportunities for the welders in marine industries. Moreover, the ever-growing applications of plastics and polymer composites have compelled these same sectors to join metals to plastics. This could pave the path for future researchers to work in this particular area and analyze all the fundamental properties of ultrasonic spot-welded joints.

Index

Printed in the United States
by Baker & Taylor Publisher Services